Inorganic Vibrational Spectroscopy

Volume 1

Inorganic Vibrational Spectroscopy

Volume 1

LLEWELLYN H. JONES

LOS ALAMOS SCIENTIFIC LABORATORY
UNIVERSITY OF CALIFORNIA
LOS ALAMOS, NEW MEXICO

1971

MARCEL DEKKER, INC., NEW YORK

MARCEL DEKKER, INC.

95 Madison Avenue, New York, New York 10016

LIBRARY OF CONGRESS CATALOG CARD NUMBER: 72-146803

ISBN NO.: 0-8247-1335-4

PRINTED IN THE UNITED STATES OF AMERICA

Preface

An understanding of the forces which hold atoms together to form molecules has been a goal of chemists for many decades. These forces arise from mutual interactions of electrons and nuclei. The vibrational spectrum of a molecule, as observed in both infrared absorption and Raman studies, is a direct manifestation of these forces. The nuclei move with approximate harmonic motion under forces determined by the electronic configuration. The vibrational spectra tell us the frequencies of the oscillations. These frequencies are related to potential energy constants whose values are determined by the electronic structure.

If one can determine for a molecule meaningful potential constants, which describe the potential field under which the nuclei move, one has a valuable quantity for correlation with electronic structure. This book is intended to outline how one can go about determining the potential constants and evaluating their significance. It is written primarily for the inorganic chemist who has not specialized in this field but wishes to apply it either to his own work or toward evaluation of other workers' results and interpretations.

The potential constants mentioned above have generally been determined as force constants. While this book was in preparation, the author was made to realize, as pointed out by J. C. Decius several years ago, that in many ways *compliance constants* are more interesting than force constants. Therefore, I have used the term potential constants, covering both force constants and compliance constants, which are, of course, related. A section on compliance constants is included.

There are a large number of molecules for which normal coordinate treatments have been reported in the literature. Unfortunately, for most of these, rather gross approximations were necessary, which make the significance of the results highly questionable. It did not seem appropriate to discuss these dubious calculations at this time; therefore, the discussion of individual molecules in this book is limited to the few for which reliable potential constants are known.

Contents

Introduction

In recent years, vibrational spectroscopy has been used extensively for studying various properties of inorganic compounds. By *vibrational spectroscopy* is meant the determination of the energy levels of the various fundamental modes of vibration occurring in the molecules or complex ions* of interest. Many of these energy levels can be determined by observing the frequencies of transitions from one level to another using infrared absorption and Raman spectroscopy. A number of commercial instruments are available for observing infrared absorption spectra; in general, at least two will be required to study the range of interest to vibrational spectroscopists, namely 50–5000 cm^{-1}. Raman spectra have come forth in much more limited volume but great strides are now being made in this direction with the availability of the laser source for Raman spectrometers.

This book will not go into experimental techniques† extensively but rather will start with the idea: "here we have the observed spectra; now what can we learn from it?" In other words, the purpose is to show what information can be obtained from the experimental data and how meaningful it is.

The most fundamental information to be obtained from vibrational spectra is a quantitative knowledge of the interatomic forces within a molecule. These forces can be treated approximately as Hooke's Law springs placed appropriately about the molecule. The force constants of these springs then determine the frequencies of the fundamental vibrations. The binding forces are determined by the electronic configuration. Thus, from the observed vibrational frequencies, we can glean information about the force constants which hopefully can then be correlated with electronic structure and bonding theories. Often it is too complex a problem to calculate force constants, but the research worker will use the magnitude of a vibrational frequency as a semiquantitative measure of the

* Henceforth the word molecules shall include complex ions.

† For experimental techniques see, for example, *An Introduction to Infrared Spectroscopy* by W. Brügel (Wiley, New York, 1962) and *Chemical Infrared Spectroscopy* by W. J. Potts, Jr. (Wiley, New York, 1963).

bond strength. For gross effects, this is often informative if used judiciously. However, the relations between frequencies and force constants are usually not simple for polyatomic molecules and the force constants are much more directly related to electronic structure.

This volume of *Inorganic Vibrational Spectroscopy* is devoted primarily to the determination of force constants from vibrational frequencies, a discussion of the meaning of force constants, and an indication of how they can best be correlated with electronic structure and bonding theories. In order to determine force constants from vibrational frequencies, it is important to use the theory of molecular vibrations as described in the literature. Though the theory will not be covered in detail in this book, the main results and their use in determining force constants will be included. Examples are given in the appendices to illustrate the procedures. Though the detail presented in these examples may seem unnecessary to some, it is felt that it will be helpful to the chemist wishing to start in the field of vibrational spectroscopy.

Though force constants have been calculated for a large number of molecules in the past few decades, most of the results are based on some very arbitrary assumptions because of lack of information. With the availability of vibrational spectra of isotopic species and other types of pertinent data, more meaningful force constants are now being reported in the literature. Many of these will be discussed in detail in this volume after the necessary procedures of calculation are outlined.

For many purposes, compliance constants are more useful than force constants. Compliance constants are elements of the inverse of the force constant matrix. Though they have received little attention in the literature, they are reported along with the force constants for a number of systems in this book. A discussion of their importance is presented in Sect. 1.13.

For many systems there is just not enough information available to calculate force constants (or compliance constants) and it may be a long time before the situation improves significantly. However, a good deal of information about stereochemistry and electronic structure can be obtained from vibrational frequencies and band intensities without force constant calculations. This arises because the characteristics of the observed spectrum, such as intensity and infrared or Raman activity, are determined by the arrangement of electrons and atoms within a molecule. Also it is often possible to obtain important evidence concerning crystal structure from infrared and Raman studies of crystalline solids, especially of single crystals. Often vibrational spectra are used for analysis for studying concentrations in equilibrium mixtures, etc. It is planned to publish later, a second volume of *Inorganic Vibrational Spectroscopy* which will be devoted to these other, less quantitative applications of vibrational spectra.

Inorganic Vibrational Spectroscopy

Volume 1

Chapter 1

Force Constants: Their Interpretation and Determination

1.1. Diatomic Molecules

Chemical bonds are found to behave like Hooke's Law springs for small displacements. Hooke's Law is given by the expression

$$F = -kX \tag{1.1-1}$$

If X is a displacement of a Hooke's Law spring from its equilibrium length, F is the force tending to restore it to equilibrium. The proportionality constant k is called the force constant. Thus, when we speak of a bond force constant, we mean the restoring force resulting from a small unit displacement from the equilibrium length. It is, therefore, also a measure of the force necessary to stretch the bond a small amount. This is an important quantity as it is determined by the electronic configuration in the molecule and thus gives information about the electronic structure and bonding. The *force constant* is a direct measure of the *bond strength* determined by the electronic distribution when the nuclei are at their equilibrium positions. Note that this is a different "bond strength" than the energy required to break a bond. More will be said about this in the next section.

A diatomic molecule acting under Hooke's Law forces is a harmonic oscillator. The potential energy (*1*) is

$$V = \tfrac{1}{2}kX^2 \tag{1.1-2}$$

and the molecule is found to vibrate with a frequency

$$\nu = (k/4\pi^2\mu)^{1/2} \tag{1.1-3}$$

The quantity X is a displacement of the bond length from its equilibrium value and μ is the reduced mass $[M_1M_2/(M_1 + M_2)]$ of the diatomic molecule. The quantum mechanical solution to the harmonic oscillator problem shows (*2*) that the vibrational energy levels can be expressed by the relation

$$G_v = \nu(v + \tfrac{1}{2}) \tag{1.1-4}$$

where $v = 0, 1, 2, \ldots$. In vibrational spectroscopy, the frequency ν and thus the energy G is generally expressed in units of waves per centimeter (cm^{-1}) which is, of course, the reciprocal of wavelength. This unit cm^{-1} is sometimes referred to as a wave number.

In reality, a diatomic molecule is not a true harmonic oscillator. Analytic functions which approximate the observed potential curves are discussed by Herzberg (*3*). These will not be discussed here, but the qualitative aspects will be brought out. As the molecule is displaced from its equilibrium atomic separation, the electrons will rearrange themselves to accommodate to the new configuration. This, in effect, changes the forces holding the atoms together. Thus, k in Eq. (1.1-2) is essentially a function of X. The potential curve near the minimum (for infinitesimal displacements) is well approximated by a parabola. Thus we may write

$$k = k_e + g(X) \tag{1.1-5}$$

$$V = \tfrac{1}{2}k_eX^2 + \tfrac{1}{2}X^2f(X) \tag{1.1-6}$$

where $f(X)$ is some function which goes to zero as X goes to zero. The equilibrium force constant, k_e, is then a measure of the strength of the bond at the minimum in the potential curve. The corresponding energies in terms of cm^{-1} of the various vibrational levels are then given by

$$G(v) = \omega_e(v + \tfrac{1}{2}) - \omega_e x_e(v + \tfrac{1}{2})^2 + \omega_e y_e(v + \tfrac{1}{2})^3 - \omega_e z_e(v + \tfrac{1}{2})^4 \quad (1.1\text{-}7)$$

By observing the fundamental and various overtone transitions, one can determine the values of ω_e (the fundamental harmonic frequency) and the various anharmonicity constants, $\omega_e x_e$, $\omega_e y_e$, etc. For example, for $H^{35}Cl$ the energy levels of the first five vibrationally excited states have been determined (*4*). From them, the data of Table 1.1-1 is calculated.* The coefficients of higher terms are essentially zero.

TABLE 1.1-1

VIBRATIONAL CONSTANTS OF $H^{35}Cl$

$\omega_e = 2991.03\ cm^{-1}$
$\omega_e x_e = 52.89\ cm^{-1}$
$\omega_e y_e = 0.247\ cm^{-1}$
$\omega_e z_e = 0.0146\ cm^{-1}$

If we knew the frequencies of only the fundamental and first overtone, we would neglect $\omega_e y_e$ and $\omega_e z_e$ and calculate $\omega_e = 2990.0$, $\omega_e x_e = 52.0$. The observed fundamental transition is 2885.9. Thus, neglect of $\omega_e y_e$ and $\omega_e z_e$ has only a small effect (0.07% in force constant) while neglect of $\omega_e x_e$ introduces a large error of about 3.5% in the harmonic frequency or 7% in the harmonic force constant. Of course, a rather large anharmonicity exists for HCl; it is generally much smaller for nonhydrogen containing molecules ($\omega_e x_e = 13.3\ cm^{-1}$ and $\omega_e y_e = +0.011\ cm^{-1}$ for CO). Thus one can obtain quite a good estimate of ω_e if only two transitions are observed and $\omega_e y_e$ is neglected. However, one must keep in mind that the only really meaningful force constant is the k_e of Eq. (1.1-5), and

$$k_e = 4\pi^2 \mu \omega_e^2 \quad (1.1\text{-}8)$$

If the anharmonic corrections are unknown, one must live with this uncertainty in the results.

In summary, then, ω_e is a hypothetical frequency for infinitesimal vibrations. It is obtained by extrapolation of the observed vibrational levels to the bottom of the potential well. It is important as data for

* These values, calculated by the author, differ slightly from those listed in (*4*).

calculation of the harmonic force constant k_e which is a measure of the forces holding the atoms together at their equilibrium length, r_e. Actually the molecule is constantly in vibration, the atoms moving apart and together about r_e. As the distance between atoms changes, the electronic configuration and thus the binding force changes also. For a measure of the binding force at the minimum of energy, which is related to the equilibrium electronic configuration, we should calculate k_e as in Eq. (1.1-8).

1.2. Bond Strength

There are various definitions of "bond strength" in the literature, as well as much use of the term without definition. In this book, bond strength

TABLE 1.2-1

Force Constants and Dissociation Energy for HI and Cl_2

	k_e (m dyn Å^{-1})	D_0 (eV)
HI	3.14	3.056
Cl_2	3.33	2.475

is used synonymously with force constant. As discussed in Sect. 1.1, this means the force necessary to stretch a bond by a small unit amount from its equilibrium length. This seems to be the best criterion for comparing the "strength" of the binding forces among different bonds.

The force constant is, of course, a different measure of strength than the dissociation energy. When a bond stretches enough to dissociate, its electronic structure (and apparent force constant) changes considerably. Usually bonds with larger force constants also have greater dissociation energies; however, there are exceptions as seen in Table 1.2-1. Thus it takes more energy to stretch a Cl—Cl bond by a small amount (compared to equilibrium bond length) than to stretch an H—I bond by the same amount; however, it takes less energy for complete separation of the two Cl atoms than for separation of the H and I atoms. By the force constant criterion, Cl_2 has a stronger bond than HI.

Cleveland has proposed (*5*) another definition of bond strength for a diatomic molecule. He feels that the most important measure of strength is the maximum force required in pulling the atoms apart. However, this maximum occurs when the molecule is stretched considerably from its equilibrium length and the electronic configuration has certainly changed. Thus, this measure of strength is not closely related to the binding forces at the equilibrium configuration. It is dependent on the anharmonicities involved. It is essentially a compromise between force constant and dissociation energy, more related to the latter.

1.3. Determination of Vibrational Frequencies of Polyatomic Molecules

The theory of molecular vibrations has been presented in the literature (*6–8*), and it will not be developed in this book. However, many of the results will be given for later application to specific molecules. In order to determine force constants for a molecule one must know most of its vibrational frequencies.

1.3.1. Vibrational Energy Levels

The vibrational potential of a polyatomic molecule is exceedingly more complicated than that of a diatomic molecule. The internal configuration for a molecule of N atoms can be defined by $3N - 6$ coordinates. Thus, the configuration of a triatomic molecule, XYZ, is completely defined by the XY distance, the YZ distance, and the XYZ angle. These distances and angles are referred to as *internal coordinates* as they define the arrangement of atoms within the molecule but say nothing about its orientation in space, which is of no concern in discussing the pure vibrations of an isolated molecule. The three coordinates mentioned above for XYZ are not unique; the XZ distance would serve in place of one of the others.

The potential energy is a function of the displacements of these internal coordinates from their equilibrium positions. Analogous to Eq. (1.1-6) for diatomic molecules, it can be written as

$$2V = \sum_{ij} F_{ij}S_iS_j + \sum_{ijk} F_{ijk}S_iS_jS_k + \cdots$$

The S_i are displacements of the various internal coordinates from their equilibrium positions. The F_{ij} are the quadratic force constants. The F_{ijk} are "cubic" constants; the cubic terms are negligible for small displacements. The potential function will be discussed further in Sect. 1.9, 1.10, and 1.12.

The forces holding the atoms together are approximately harmonic, resulting in normal vibrations of $n = 3N - 6$ nearly harmonic oscillators ($n = 3N - 5$ for linear molecules). The vibrational energy to second order in vibrational quantum number is given by Eq. (1.3-1).

$$G_{v_1,v_2,\ldots} = \sum_i \omega_i(v_i + \tfrac{1}{2}d_i) + \sum_i \sum_{j\geq i} X_{ij}(v_i + \tfrac{1}{2}d_i)(v_j + \tfrac{1}{2}d_j) + W(l_i, l_j, \ldots) \tag{1.3-1}$$

In Eq. (1.3-1) the ω_i are the fundamental vibrational frequencies at infinitesimal amplitude (they are actually extrapolations from observed frequencies to the values at infinitesimal amplitude). Thus they are the harmonic oscillator frequencies which would arise from the forces imposed by the equilibrium electronic configuration. They are referred to in this book as the harmonic frequencies. The d_i are the degeneracies of the frequencies ω_i. For nonlinear molecules, $\sum_i d_i = 3N - 6$; for linear molecules, $\sum_i d_i = 3N - 5$. The X_{ij} are anharmonic constants arising primarily from cubic and higher terms in the potential function. They correspond to the $\omega_e x_e$ of diatomic molecules. The v_i are the quantum numbers of the various vibrational modes in energy level $G_{v_1,v_2,\ldots}$. The term W is a function of the internal angular momentum quantum numbers l_i, l_j, etc. It is zero for nondegenerate vibrations. The quantities l_i may have the values v_i, $v_i - 2, \ldots$ 1, or 0 for degenerate vibrations.

The first sum on the right in Eq. (1.3-1) is that derived from a quadratic potential function, for which the normal vibrations are true harmonic oscillations. If cubic and higher terms in the potential function are significant, we must include the second sum of anharmonic terms (including W) to reproduce the observed energy level pattern. Higher terms, such as those involving $(v_i + \frac{1}{2}d_i)(v_j + \frac{1}{2}d_j)(v_k + \frac{1}{2}d_k)$, are expected to be very small as found for diatomic molecules [see Eq. (1.1-7) and Table 1.1-1]. This is born out in practice for the few cases where evidence is available. If a sufficient number of combination and overtone transitions are observed, for which $\sum_i \Delta v_i > 1$, it is possible to calculate the anharmonic corrections, X_{ij} and W of Eq. (1.3-1). Illustration of this procedure is given in Appendix I. In most cases, most of the anharmonic constants are unknown and therefore neglected. The calculated force constants are then only an approximation to the true harmonic force constants and must be treated as such. In a few cases, a sufficient number of vibrational levels has been determined to calculate many or all of the anharmonic corrections, X_{ij} and W, of Eq. (1.3-1). In later chapters, specific cases will be discussed.

Another factor which must be considered in evaluating the energy levels of polyatomic molecules is Fermi resonance. It may happen that two vibrational levels belonging to different vibrations (or combinations of vibrations) will have nearly the same energy. If these two levels have the same symmetry, they may mix, resulting in a repulsion of the two levels and a sharing of intensity for infrared absorption or Raman emission. The classic example is CO_2 for which the levels 02^00* and 100 both come at about 1335 cm^{-1} above the ground vibrational state, 0 0 0. Since these levels have the same symmetry $\sum_g^+$, they interact and the resulting Raman lines for both $2\nu_2$ and ν_1 appear strongly at 1388.3 and 1285.5 cm^{-1}. Such a perturbation must be accounted for in applying Eq. (1.3-1); thus, the energies, $G_{v_1,v_2,\ldots}$, of this expression apply for the levels unperturbed by Fermi resonance. This subject is treated in the literature (see, for example, (7) p. 215 ff.).

1.3.2. Vibrational Transitions

A schematic representation of the energy levels of a triatomic molecule is given in Fig. 1.3-1. The quantum numbers of the vibrational levels are

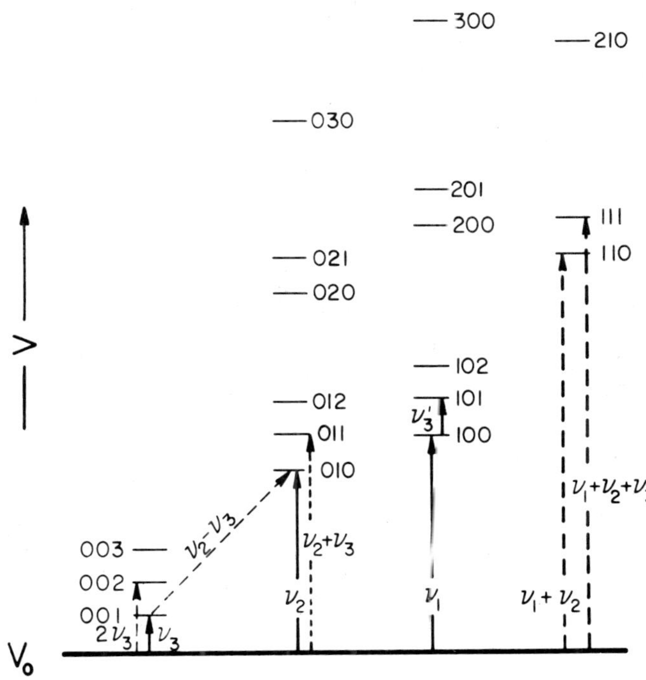

Fig. 1.3-1 Schematic vibrational energy diagram for a triatomic molecule.

* The superscript ^{0} on the 2 is the value of l in Eq. (1.3-1).

expressed as v_1, v_2, v_3; thus, the notation 1 0 0 means $v_1 = 1$, $v_2 = 0$, $v_3 = 0$ and 0 2 1 means $v_1 = 0$, $v_2 = 2$, $v_3 = 1$. In the harmonic oscillator approximation, the levels successive for $\Delta v_k = 1$ (such as 0 0 1, 0 0 2, 0 0 3, . . .) are equally spaced. However, in practice there is generally a slight convergence toward higher v_i.

First let us talk about absorption of radiation. This involves a transition from one vibrational level (usually 0 0 0) to a higher vibrational level. For example, if we observe infrared absorption corresponding to the transition from 0 0 0 to 0 0 1, we call it ν_3 and it tells us the energy which 0 0 1 has above the ground state, 0 0 0. From Eq. (1.3-1) this then gives us an approximate value for the vibrational frequency, ω_3.

The transitions $000 \rightarrow 001$, $000 \rightarrow 010$, and $000 \rightarrow 100$, for which $\Delta v_i = 1$, are called *fundamental* transitions. The frequencies ν_1, ν_2, and ν_3 associated with these transitions are then *fundamental frequencies.* The frequency ν_3', arising from the transition $100 \rightarrow 101$, is also a fundamental frequency but is called a *hot* transition, giving rise to a *hot band** in the absorption spectrum. The absorption frequency ν_3' would be exactly at the position of ν_3 for true harmonic oscillations; however, in practice it is often displaced slightly because of anharmonicity.

The dotted line transitions, such as $\nu_2 + \nu_3$, are called binary combinations. A binary combination such as $2\nu_3$ is also called the first overtone ($3\nu_3$ is the second overtone, etc.). The binary combination $\nu_2 - \nu_3$ is called a *difference* band. It is also a *hot* band.

In the Raman effect, the process is more complex than for absorption spectra. A detailed discussion is given by Evans (*9*). In brief, the molecules are bombarded with photons which are then scattered. During the scattering process, some of the photons excite some of the molecules to higher energy levels. The loss of energy of the photons then equals the gain in energy of the molecule and we observe emission at a frequency lower than the incident light by the transition energy. The observation of Raman shifts can arise from vibrational transitions such as discussed above for infrared absorption spectra, and Fig. 1.3-1 applies to Raman spectra also. In the case of Raman spectra, the reverse transitions (anti-Stokes) take place also; that is, the incident photons can extract vibrational energy from the molecules by inducing transitions such as $100 \rightarrow 000$.

* This is called a *hot band* because it starts in an excited state. By cooling the material one can reduce the number of molecules in this excited state and thus reduce the intensity of the hot band.

1.3.3. Infrared Activity

The absorption of electromagnetic radiation by a molecule so as to excite it from one vibrational state to another vibrational state involves the interaction of the oscillating electric vector of the radiation with what is classically considered as an oscillating dipole in the molecule. During a normal vibration of a molecule, the dipole moment can be expressed to the first order as

$$\mu^{(k)} = \mu_0 + \left(\frac{d\mu}{dQ_k}\right) \Delta Q_k \qquad (1.3\text{-}2)$$

where μ_0 is the permanent dipole moment and Q is a normal coordinate describing the motion of the molecule. The coefficient $d\mu/dQ_k$ tells how much the dipole moment changes during the vibration. If $d\mu/dQ_k$ is zero, $\mu^{(k)}$ equals μ_0 and no infrared absorption can occur. If any one of the components, $d\mu_x/dQ_k$, $d\mu_y/dQ_k$, or $d\mu_z/dQ_k$, is different from zero, the vibration is infrared-active. If $d\mu_\alpha/dQ_k$ is different from zero, the vibration ν_k must fall under the irreducible representation containing the translation, T_α ($\alpha = x, y, z$). Derivation of these selection rules has been presented by several authors. The reader is referred especially to the book *Molecular Vibrations* by Wilson et al. (*6*).

To determine which vibrations are infrared-active, one can look at the character table (Appendix IX) for the point group in question and note which irreducible representations contain the translations T_x, T_y, T_z (or x, y, z). All vibrations in these representations can give rise to absorption of radiation (they are infrared-active). All other vibrations are not infrared-active. Section 1.4 discusses the classification of vibrations under irreducible representations.

Even though a vibration is infrared-active, it may not be observed in absorption. It has been shown (*6*) that the intensity of an absorption band is proportional to

$$\left(\frac{\partial\mu_x}{\partial Q_k}\right)^2 + \left(\frac{\partial\mu_y}{\partial Q_k}\right)^2 + \left(\frac{\partial\mu_z}{\partial Q_k}\right)^2$$

If, for example, for the vibration ν_k,

$$\frac{\partial\mu_x}{\partial Q_k} = \frac{\partial\mu_y}{\partial Q_k} = 0$$

and $\partial\mu_z/\partial Q_k$ is very small, the band may be very weak. The intensity of absorption is a strong function of the electronic rearrangement during the vibration.

For a true harmonic oscillator, the only vibrational transitions which can occur are for $\Delta v = \pm 1$; thus, strictly speaking only the fundamental transitions are infrared-active. However, we do observe overtones and combinations in absorption, generally quite a bit weaker than the fundamentals. The observation of overtones and combinations is strong evidence that real molecules are not true harmonic oscillators.

The selection rules based on symmetry apply also to overtones and combination bands. Overtones and combinations often consist of more than one component, usually falling under different representations. The representations contained by a given overtone or combination are given by Herzberg [(*7*), pp. 126–129]. As an example, consider tetrahedral CH_4. The vibration ν_3 is of F_2 symmetry and is, thus, infrared-active. The first overtone $2\nu_3$ contains components $A_1 + E + F_2$; thus, it has only one infrared-active component as A_1 and E are not infrared-active. It has three Raman-active components as A_1, E, and F_2 are all Raman-active.

1.3.4. Raman Activity

When molecules are bombarded with electromagnetic radiation, a portion of this radiation interacts with the electrons of the molecule and is scattered in all directions. Most of this scattered radiation has the same frequency as the incident radiation (Rayleigh scattering); however, a portion has frequencies different from that of the incident radiation by plus or minus some of the various vibrational or rotational energies or both. Thus, if the incident radiation is the 4358 Å line of mercury (energy about 22,946 cm^{-1}), the cyanide ion, which has a vibrational frequency of 2080 cm^{-1}, will show Raman peaks at 20,866 cm^{-1} and 25,026 cm^{-1}. The lower energy peak (Stokes line) will be considerably more intense than the higher energy peak (anti-Stokes line) because it originates in the ground vibrational level, while the anti-Stokes line originates in an excited vibrational level (which is less populated at ordinary temperatures). The intensity of Raman scattering varies as ν^4, thus higher frequency incident light gives much stronger Raman lines, other factors being equal.

The interaction of the electric vector of the incident light with the molecule to produce scattering occurs because the electric vector induces a dipole in the molecule. This dipole is induced because the molecule is polarizable. For a molecule executing small vibrations, the polarizability can be expressed as

$$\alpha = \alpha^0 + \sum_k \left(\frac{\partial\alpha}{\partial Q_k}\right)_0 \Delta Q_k + \text{higher terms} \tag{1.3-3}$$

analogous to Eq. (1.3-2) for the permanent dipole moment.

The quantity α^0 refers to the polarizability of the nonvibrating molecule whereas $(\partial\alpha/\partial Q_k)$ is the change in polarizability accompanying the motions of normal vibration k. The intensity of the Raman scattering for vibrational frequency ν_k is proportional to $(\partial\alpha/\partial Q_k)^2$.

If $(\partial\alpha/\partial Q_k)$ is zero, no Raman peak can occur for ν_k. Let us consider the polarizability a little further. For the stationary molecule, the Cartesian axes x, y, and z are defined. When plane polarized radiation with electric vector E strikes the molecule, it can be resolved into components E_x, E_y, and E_z along the three axes. The induced dipole moment P is not necessarily parallel to E and can also be resolved into three components, P_x, P_y, and P_z. Since P_x will contain components from each of the three components of E, we write

$$P_x = \alpha_{xx}E_x + \alpha_{xy}E_y + \alpha_{xz}E_z$$

and similarly

$$\begin{aligned} P_y &= \alpha_{yx}E_x + \alpha_{yy}E_y + \alpha_{yz}E_z \\ P_z &= \alpha_{zx}E_x + \alpha_{zy}E_y + \alpha_{zz}E_z \end{aligned} \tag{1.3-4}$$

In these relations $\alpha_{ij} = \alpha_{ji}$. The square matrix of the coefficients α_{ij} is known as the *polarizability tensor*. For Raman scattering, we are interested only in the changes in polarizability, $\partial\alpha/\partial Q_k$, accompanying normal vibrations. In this case, the terms of the polarizability tensor are the derivatives $(\partial\alpha_{xx}/\partial Q_k)_0$, $(\partial\alpha_{xy}/\partial Q_k)_0$, etc. These are abbreviated as α'_{xx}, α'_{xy}, etc., and they transform under symmetry operations as the corresponding quadratic functions of Cartesian translations $(T_x)^2$, T_xT_y, etc. This leads to the conclusion that only vibrations which are under irreducible representations transforming as one ore more of the quantities α'_{xx}, α'_{yy}, α'_{zz}, α'_{xy}, α'_{xz}, and α'_{yz} are Raman-active. Thus, one can inspect the character tables as in Appendix IX and note which irreducible representations contain the quantities α'_{xx}, α'_{xy}, etc. In these tables, transformations of x^2, xy, etc., are given; these correspond to α'_{xx}, α'_{xy}, etc., respectively. These representations are Raman-active, meaning they can give rise to peaks in the Raman spectra. The intensity may be quite weak depending on the magnitude of the α_{ij}. All vibrations in representations not containing at least one of the α_{ij} are Raman inactive.

A further important characteristic of Raman peaks is their *depolarization ratio*. The theory is presented nicely by Wilson et al. [(*6*), pp. 43–47]. The usual mercury arc source experimental setup gives irradiation of a cylindrical sample tube by a helical light source whose axis is coincident with the axis of the sample tube. The Raman scattering is observed along the axis of the sample tube. If a cylinder of Polaroid is placed concentrically between the sample tube and the light source, the Raman lines show a difference in intensity depending on which way the Polaroid is oriented. If we note the intensity as $I_\perp$ when the Polaroid passes light polarized perpendicular to the sample tube axis, and $I_\parallel$ when the Polaroid passes light polarized parallel to the sample tube axis, the *depolarization* ratio is

$$\rho = I_\parallel / I_\perp$$

Theoretically the value of ρ will be $\frac{6}{7}$ for nontotally symmetric irreducible representations. Such Raman lines are said to be *depolarized.* For the totally symmetric representation (in which no symmetry elements are destroyed during the vibration), the value of ρ may be anywhere from zero to $\frac{6}{7}$ depending on the values of the α'_{ij}. For cubic point groups such as T_d and O_h, ρ is zero for totally symmetric vibrations. The actual experimental results generally give a value of ρ somewhat higher than the theoretical one, because of various experimental factors. Generally ρ is sufficiently less than $\frac{6}{7}$ for totally symmetric vibrations so that its observation is of great help in assigning the observed Raman lines to the appropriate symmetry species (irreducible representation).

With the laser source, the usual arrangement is somewhat different. The incident light is polarized, but also the Raman emission is analyzed with a polarizer. The theoretical depolarization ratio for the appropriate configurations is zero to $\frac{3}{4}$ rather than zero to $\frac{6}{7}$.

1.3.5. Assignment of Frequencies

From a thorough study of the infrared and Raman spectra of a molecule in one, or preferably more, of the available phases (gas, liquid, solution, or solid), one can generally observe many, if not all, of the $3N - 6$ fundamental vibrational frequencies. Often a fundamental frequency will not appear by itself in the spectrum but will appear as a combination band, arising from two or more simultaneous vibrational transitions. Thus, in favorable cases for a molecule having n distinct normal vibrations, one can observe n frequencies associated with fundamental transitions. There remains the problem of how to assign these n frequencies. There are $n!$ different ways of making the assignment. For example, consider a bent triatomic molecule such as ClNO. There will be an NO stretching frequency (ν_{NO}), an NCl stretching frequency (ν_{NCl}), and a ClNO bending frequency (δ_{ClNO}). Three fundamentals have been observed, at 1799, 592, and 332 cm^{-1}. Without any a priori information, we have $3! = 6$ choices of assignments as shown in Table 1.3-1.

The symbol ν here represents a stretching frequency while δ represents a bending frequency. Actually the vibrations are not totally in one coordinate. Thus, the NO stretching vibration probably involves a little NCl stretching and a little ClNO bending. The closer the frequencies are, the more likely it is that the vibrations intermix the coordinates. Nevertheless, except in rare cases, a given vibration involves primarily one type of coordinate.

TABLE 1.3-1

POSSIBLE FREQUENCY ASSIGNMENTS[a] FOR ClNO

ν_{NO}	1799	1799	592	592	332	332
ν_{NCl}	592	332	1799	332	1799	592
δ_{ClNO}	332	592	332	1799	592	1799

[a] Values are expressed in cm^{-1}.

For ClNO, with neglect of the interaction constants, we find six different force constant solutions. As discussed in Sect. 1.10, for each force constant solution we can calculate the relative amplitudes and potential energy distribution of the three internal coordinates (ΔR_{NO}, ΔR_{NCl}, and $\Delta\alpha_{ClNO}$) for each observed frequency. Six different force constant solutions are associated with the six different frequency assignments given above.

The concept of *group frequencies* is quite useful in deciding upon the correct frequency assignment. This concept is simply that a given chemical entity will have roughly the same frequency in different compounds. It is quite useful when the frequency of the group is well separated from frequencies arising from other groups. For example, a CH stretching vibration occurs in the neighborhood of 3000 cm^{-1}, a C≡N stretch occurs around 2050 to 2200 cm^{-1}, a C≡O stretch around 1950 to 2150 cm^{-1}, etc. Some very useful group frequency charts are given by Nakamoto (*8*).

The vibrational frequency of free NO is about 1876 cm^{-1}. In ClNO, the bond length is about the same as in free NO, so we expect the NO stretching frequency to be not greatly different. This means that 1799 cm^{-1} must be the NO stretching frequency for ClNO. Thus, the last four assignments of Table 1.3-1 can be thrown out. We would need further information to choose between the first two assignments. Usually stretching vibrations occur at higher frequencies than bending; however, this is not always true. A study of isotopically substituted species in this case has given a unique assignment as shown in Sect. 2.13. Each of the three frequencies exhibit a different percentage isotopic shift. The observed shift when compared with the calculated shift for the two possible assignments yields the correct force constant solution and frequency assignment. A similar procedure has been used (*9a*) to remove ambiguity in the AuC stretching and AuCN bending assignments of $Au(CN)_2Cl_2^-$. This was based on spectra of ^{15}N and ^{13}C substituted species.

From the shape of vibration–rotation bands in gas phase studies, one can often rule out certain frequency assignments. For symmetric top molecules, parallel bands (vibrational dipole change parallel to the unique axis) will generally have different shapes and structure than perpendicular bands. A detailed discussion has been given by Herzberg (*7*). Of course, the determination of infrared and Raman activity and Raman depolarization ratios are important means of frequency assignment. Polarized infrared and Raman studies of single crystals are generally quite useful in assigning frequencies. The assignment of frequencies will be discussed further in Vol. II.

1.4. Symmetry of Normal Vibrations

In order to carry the discussion of molecular vibrations further, it is fruitful to consider the symmetry of molecules. Great simplification is obtained when molecular symmetry is utilized in the determination of relations between vibrational frequencies and force constants. Symmetry information is also of great value for determination of the number and degeneracies of fundamental vibrations, the selection rules for infrared and Raman spectra, the number of independent force constants, etc. Many treatments of molecular symmetry exist in the literature (*6–11*); this book will not try to cover the theory but rather will present the results necessary for application to vibrational spectroscopy. For more thorough discussions of symmetry, the reader is referred especially to (*6*), (*10*), and (*11*).

1.4.1. Symmetry Elements and Symmetry Operations

A molecule can be classified as belonging to a certain *point group* depending on the *symmetry elements* it possesses. The possible symmetry elements are: C_n an n-fold rotational axis of symmetry; σ, a plane of symmetry; i, a center of symmetry; and S_n, an n-fold rotation-reflection axis.

A *symmetry element* is best defined by considering a *symmetry operation*. Thus, if we rotate a molecule by $2\pi/n$ rad around an n-fold axis of symmetry, C_n, each atom, considered as a point, will either remain where it was before the operation or reside at a position formerly occupied by an identical atom. The symmetry operation performed in this case is labeled C_n^1. The operation C_n^k is a rotation of $2\pi k/n$ rad about the symmetry element, C_n.

If a molecule possesses a plane of symmetry, it means that for each atom at a perpendicular distance r from the plane there is an identical atom at the same distance r on the opposite side of the plane in the same perpendicular line. A plane of symmetry perpendicular to the highest fold symmetry axis (the unique axis) is called a horizontal plane of symmetry σ_h. A plane of symmetry which includes the unique axis of symmetry is called a vertical plane of symmetry σ_v. A vertical plane of symmetry which also bisects the angle between two C_2 axes is, however, generally referred to as σ_d. The operation of interchanging all atoms on one side of a plane with identical atoms on the other side is called a reflection through the plane.

If a point exists such that for each atom of a molecule a straight line through the atom and this point meets an identical atom at the same distance on the other side of the point, this point is a center of symmetry i.

If a rotation of C_n is performed about an axis, followed by a reflection through a plane perpendicular to this axis, and all atoms remain either at their original position or at positions previously occupied by identical atoms, the molecule possesses a rotation-reflection axis, or "improper rotation" axis, S_n.

Two symmetry elements in a molecule often imply a third element. For example, if a molecule has a rotation axis, C_n, and a horizontal plane of symmetry, σ_h, it must also possess a rotation-reflection axis, S_n. The operation S_2 is the same as i.

Another class of operation, the identity E is also important in the mathematical relations of molecular symmetry. This is the operation of leaving the molecule in its original position.

1.4.2. Point Groups

The symmetry elements contained in a molecule are specified when its *point group* is given. A brief description of point groups is given in Table

1.4-1. The symmetry elements given are those required besides the "identity." More elements may exist in some cases unless "only" is stated.

By inspection of the molecular structure, known or assumed, one can determine the symmetry elements present and, thus, the point group. The character tables for the various point groups of interest are given in Appendix IX. The first column in the character table lists the irreducible representations which classify the vibrations for that point group.

TABLE 1.4-1

DESCRIPTION OF POINT GROUPS

Point group	Necessary symmetry elements
C_1	No symmetry
C_i	Center of symmetry only
C_s	Plane of symmetry only
C_n	n-fold axis of symmetry only
C_{nv}	n-fold axis and $n\sigma_v$ only
C_{nh}	n-fold axis and σ_h
D_n	n-fold axis and n 2-fold axes perpendicular to C_n
D_{nd}	D_n with n-vertical planes of symmetry, σ_d
D_{nh}	D_n with σ_h
$C_{\infty v}$	Linear with no center of symmetry
$D_{\infty h}$	Linear with center of symmetry, i
S_n	$C_{n/2}$ and S_n
T_d	Tetrahedral (see Character Table, Appendix IX)
O_h	Octahedral (see Character Table)

Vibrations which fall under an E representation are doubly degenerate (two vibrations of the same frequency). F representations are triply degenerate, while A and B are nondegenerate. For the theory of character and representations, the reader is referred to (*6*) and (*10*). We will now describe a method for determining the number of vibrations in each representation for a molecule of known point group, as outlined in (*6*).

1.4.3. Symmetry Classification of Fundamental Vibrations

After determining the various symmetry elements and thus the point groups of a molecule, it is possible to evaluate the number of vibrational modes for a given irreducible representation of the molecule. It is a matter of substitution in the equation

$$n_\gamma = (1/h) \sum_R g_R \chi_R^\gamma \chi_R \tag{1.4-1}$$

where

χ_R^γ is the character of irreducible representation γ for operation R as given in the Character Tables (Appendix IX),

χ_R is the character for the transformation representing symmetry operation R (this will be explained below),

g_R is the number of operations in a given class; it is the numerical coefficient of the operation symbol in the top row of the character table,

h is the total number of operations in the group ($\sum_R g_R$),

n_γ is the number of vibrations in irreducible representation γ, counting all the degenerate components for E and F representations.

TABLE 1.4-2[a]

CONTRIBUTION TO CHARACTER PER UNSHIFTED ATOM

Proper Operations		Improper Operations	
R	χ_R	R	χ_R
C_n^k	$1 + 2\cos(2\pi k/n)$	S_n^k	$-1 + 2\cos(2\pi k/n)$
$E = C_1^k$	3	$\sigma = S_1^1$	1
C_2^1	-1	$i = S_2^1$	-3
C_3^1, C_3^2	0	S_3^1, S_3^2	-2
C_4^1, C_4^3	1	S_4^1, S_4^3	-1
C_6^1, C_6^5	2	S_6^1, S_6^5	0

[a] Reprinted from (*6*), p. 105, by courtesy of McGraw-Hill, New York.

We shall now define the character of the transformation R, χ_R. If a given atom is shifted when operation R is carried out, it contributes zero to the character. If an atom is unshifted, it contributes $1 + 2\cos(2\pi k/n)$ for a rotation of $2\pi k/n$ rad, noted as C_n^k. The contribution per unshifted atom for a rotation-reflection of $2\pi k/n$ rad is $-1 + 2\cos(2\pi k/n)$. Thus, for E, $2\pi k/n = 0$ and $\chi_R = 3$ for each unshifted atom. For C_2^1, a rotation of π about C_2, $\chi_R = -1$. The operation σ is a rotation-reflection with $2\pi k/n = 0$. The χ_R for various operations are given in Table 1.4-2.

Thus, from the character tables (Appendix IX and Table 1.4-2) and Eq. (1.4-1) we can determine the number of vibrations in the various irreducible representations.* This number includes the nongenuine

* One can also determine the number of vibrations in each species by using the tables in (*7*), pp. 134–139 or (*12*), pp. 44–50.

vibrations of translation and rotation. They may be subtracted off, as their species are given in the character tables.

If we are only interested in the genuine vibrations, the problem is simpler. We select the required number of internal coordinates and note how these permute among themselves when a symmetry operation is performed. If a given internal coordinate is shifted to a new position, it contributes zero to the character. If a coordinate is unchanged by the operation, $\chi_R = 1$. If it is changed only in sign, $\chi_R = -1$. We still use Eq. (1.4-1). An example is given in Appendix II. The results are more specific as they tell us not only how many vibrations there are in a given irreducible representation but also what internal coordinates are involved.

1.5. Symmetry Coordinates

In deriving relations between vibrational frequencies and force constants, it is highly advantageous to work with internal symmetry coordinates. An internal symmetry coordinate is a linear combination of internal coordinates which transforms according to a given irreducible representation. When the various symmetry operations are performed on the molecule, a symmetry coordinate will transform according to the character of its irreducible representation. A convenient equation [(*6*), p. 119] for deriving symmetry coordinates for nondegenerate representations is

$$S_i^\gamma = N \sum_R \chi_R^\gamma R(S_t) \qquad (1.5\text{-}1)$$

In this equation $R(S_t)$ stands for the coordinate to which the displacement of internal coordinate S_t is transferred by the operation R, taken with the appropriate sign. χ_R^γ is the character for operation R under representation γ as given in the character tables. N is a normalization constant chosen so that the sum of the squares of the coefficients of the various internal coordinates in the expression for S_i^γ is equal to one. The summation is over all symmetry operations R. S_i^γ is the resulting symmetry coordinate.

For degenerate representations, Eq. (1.5-1) can be applied to any one internal coordinate, S_t, but other coordinate sets, S_u, S_v, etc., must then be properly "oriented"; that is, a linear combination must be chosen so as to exhibit the same symmetry properties as the S_t used for the generation of the first set. An example of the derivation of symmetry coordinates is given in Appendix II for $M(CN)_4^{-x}$ of D_{4h} symmetry.

For molecules in which a given internal coordinate set contributes more than one vibration to a given irreducible representation ($n_{\gamma,i} > 1$), special considerations are required. The reader is referred to (*6*), p. 135, for a description of a satisfactory procedure for deriving suitable symmetry coordinates for such a case. Often the appropriate symmetry coordinates are obvious so they can be written down without further consideration.

After obtaining the various symmetry coordinates, one should check to be sure that they are all orthogonal and normalized. Thus, if the symmetry coordinates are written as

$$S_k^\gamma = \sum_t a_{t,k}^\gamma S_t$$

in which the t denotes the various internal coordinates, and the $a_{t,k}^\gamma$ are numerical coefficients, the orthogonality and normalization requires that $\sum_t a_{t,k}^\gamma a_{t,l}^{\gamma'} = \delta_{kl}\delta_{\gamma\gamma'}$ in which δ_{ij}, the Kronecker delta symbol, equals unity if $i = j$ and is zero otherwise.

1.6. Determination of Relations between Force Constants and Vibrational Frequencies

By considering the equations of motion for the atoms of a molecule, assuming a potential function analogous to Eq. (1.1-2) for diatomic molecules, we can derive relations among the frequencies of vibration and the force constants. Then, after observation of the frequencies, we can make some calculations and learn something about the interatomic forces. As an example of such a procedure, in Appendix III the treatment of linear *ABC* is carried out starting with a quadratic potential function, Cartesian coordinates, and LaGrange's equations of motion. It is hoped that this procedure will give the unfamiliar reader a feeling for the theory of normal vibrations. This is referred to as a Normal Coordinate Analysis as from the solution one can determine a Normal Coordinate for each vibration, which tells how each atom moves during that mode of vibration.

Much better methods exist for calculating these relations, using the *F* and *G* matrices to be discussed presently; however, the results depend on the same basic principles. A treatment of more complicated molecules, using Cartesian coordinates as in Appendix IIIa, becomes extremely laborious. Therefore, we resort to the use of symmetry coordinates and the *F* and *G* matrices as described below. After obtaining the symmetry coordinates for a molecule, as in Sect. 1.5, it is possible to obtain relations between the vibrational frequencies and the molecular force constants.

The theory is developed in (*6*) and only the results and their use will be discussed here.

As shown by Wilson et al. [(*6*), p. 309], the secular equation in terms of the F and G matrices is readily derived in matrix notation. In the quadratic approximation, the potential energy is expressed as $2V = S'FS$ and the kinetic energy as $2T = \dot{S}'G^{-1}\dot{S}$. F is the force constant matrix, S is a matrix of internal coordinates, and G^{-1} is called the kinetic energy matrix. It has units of mass if the S matrix has units of length. The prime indicates the matrix transpose. In terms of the normal coordinates, Q, we write

$$2V = Q'\Lambda Q \qquad \text{and} \qquad 2T = \dot{Q}'Q$$

Λ is a diagonal matrix of the frequency parameters, $4\pi^2\nu_i^2$, where ν_i is the vibrational frequency. The normal coordinates Q are related to the internal coordinates, S, by the L matrix

$$S = LQ$$

By substituting LQ for S in the equations for V and T, we find that $L'FL = \Lambda$ and $L'G^{-1}L = E$. By simple matrix manipulation [(*6*), p. 309], we arrive at two forms of the secular equations

$$GFL = L\Lambda \tag{1.6-1}$$

$$FG(L^{-1})' = (L^{-1})'\Lambda \tag{1.6-2}$$

The solutions for these two equations are

$$|GF - E\Lambda| = 0 \tag{1.6-3}$$

with L as eigenvector, and

$$|FG - E\Lambda| = 0 \tag{1.6-4}$$

with $(L^{-1})'$ as the eigenvector. These determinants are called the secular determinants. The G matrix is defined in (*6*) p. 61, and later in this section.

The same relations can be arrived at by using the expanded forms of the potential and kinetic energy.

$$2V = \sum_{i,j} F_{ij}S_iS_j$$

$$2T = \sum_{i,j} G_{ij}^{-1}\dot{S}_i\dot{S}_j$$

Application of LaGrange's Equations of Motion,

$$\frac{d}{dt}\left(\frac{\partial T}{\partial \dot{S}_i}\right) + \frac{\partial V}{\partial S_i} = 0$$

as in Appendix III, yields

$$\sum_j (F_{ij} - G_{ij}^{-1}\lambda)S_j = 0, \qquad i = 1, n \tag{1.6-5}$$

This set of equations makes up the secular determinant expressed in matrix notation as $|F - G^{-1}\Lambda| = 0$. As is also shown in (*6*), if the determinant of coefficients of S_j in (1.6-5) is multiplied by the determinant of G, we arrive at the secular determinants (1.6-3) and (1.6-4), for which λ appears only in the diagonal elements.

Each symmetry block (irreducible representation) can be treated separately as there are no cross terms between different symmetry blocks. As an illustration of the great simplification achieved by the use of symmetry coordinates, consider the $M(CN)_4^{-x}$ ion treated in Appendix II. There are sixteen normal modes of vibration of different frequencies (degenerate vibrations counted only once). This would result in a sixteen by sixteen secular determinant. However, consideration of symmetry reduces the problem to solution of one 4 × 4, five 2 × 2, and two 1 × 1 determinants.

For each symmetry block we wish to determine a G matrix and an F matrix. The former is the "inverse kinetic energy" matrix and the latter is made up of the symmetry force constants. In the secular equation for an $n \times n$ symmetry block there are $n + 1$ terms starting with λ^n and containing terms in decreasing powers of λ down to zero. The expanded form of the secular equation is obtained by expanding either (1.6-3) or (1.6-4). The general term is expressed as

$$(-1)^s \lambda^{n-s} \Sigma G^{(s)} F^{(s)} \tag{1.6-6}$$

where $G^{(s)}$ is any s-rowed minor of the determinant of the G matrix, $F^{(s)}$ is the corresponding s-rowed minor of the determinant of the F matrix, and the sum is over all possible such s-rowed minors. s runs from 0 to n, λ is the eigenvalue, $4\pi^2\nu^2$, where ν is the vibrational frequency.

The G matrix is a function of masses and atomic distances; thus, it can be determined if the molecular geometry is known. The simplest procedure is to write the symmetry coordinates in terms of Cartesian displacement coordinates obtained by transformation of the internal displacement coordinates.

$$S_t^\gamma = \sum_{i=1}^{3N} B_{ti}^\gamma \xi_i \tag{1.6-7}$$

for each symmetry coordinate in the symmetry block. In this equation ξ_i is a unit displacement of a given atom along a given Cartesian axis.

B^{γ}_{ti} is the coefficient of ξ_i which satisfies the relation for the symmetry coordinate, S^{γ}_{t}, in terms of internal coordinates of type t. After determining the B^{γ} matrix, it is possible to calculate the G^{γ} matrix from

$$G^{\gamma}_{tt'} = \sum_{i=1}^{3N} \frac{1}{m_i} B^{\gamma}_{ti} B^{\gamma}_{t'i} \tag{1.6-8}$$

in which m_i is the mass of atom i. The subscripts t and t' run from 1 to n for a symmetry block containing n frequencies. The G matrix is obviously symmetric, $G^{\gamma}_{tt'} = G^{\gamma}_{t't}$. As an example, the G matrices for $M(CN)_4^{-x}$ are derived in Appendix II.

The F matrix to be used in Eq. (1.6-6) is the symmetry force constant matrix defined by the potential energy expression

$$2V = \sum_{\gamma} \sum_{tt'} F^{\gamma}_{tt'} S^{\gamma}_{t} S^{\gamma}_{t'} \tag{1.6-9}$$

These S^{γ}_{t} are the same symmetry coordinates as used in Eq. (1.6-7).

The symmetry force constants $F^{\gamma}_{tt'}$ are linear combinations of the force constants for individual internal coordinates, including constants for interaction among these coordinates.

To determine the symmetry force constant in terms of force constants for the individual internal coordinates, one inspects the symmetry coordinate

$$S^{\gamma}_{t} = \sum a^{\gamma}_{it} S_i \tag{1.6-10}$$

The S_i are internal coordinates such as changes in bond lengths or angles. The internal coordinate force constant matrix is

	S_1	S_2	S_3	—
S_1	F_{S_1}	$F_{S_1S_2}$	$F_{S_1S_3}$	—
S_2	$F_{S_1S_2}$	F_{S_2}	$F_{S_2S_3}$	—
S_3	$F_{S_1S_3}$	$F_{S_2S_3}$	F_{S_3}	—

The diagonal terms, which are the primary force constants, are given with only one subscript for clear distinction from the off-diagonal terms. The off-diagonal force constants are interaction constants. Their physical significance will be discussed in a later section.

To determine the symmetry force constant for a given symmetry coordinate, one can use the relation

$$F^{\gamma}_{t} = \left(\sum_{i} a^{\gamma}_{it} F_{S_i S_k} \right) \bigg/ (a^{\gamma}_{kt}) \tag{1.6-11}$$

where a_{kt}^{γ} is the coefficient of an arbitrarily selected internal coordinate, S_k. $F_{S_kS_k}$ we write as F_{S_k} as mentioned above. For the off-diagonal symmetry coordinates, the expression is slightly altered to

$$F_{tt'}^{\gamma} = \left(\sum_{i} a_{it}^{\gamma} F_{S_iS_k}\right) \Big/ (a_{kt'}^{\gamma}) \tag{1.6-12}$$

The number of symmetry force constants (neglecting redundant coordinates) is equal to the number of different independent internal coordinate force constants. If the symmetry force constants can be determined, the independent internal coordinate force constants can be calculated therefrom, using the relations (1.6-11) and (1.6-12).

1.7. Solution of the Secular Equation

The general term of the secular equation is given by (1.6-6) so the secular equation can be written as

$$\sum_{s=0}^{n} (-1)^s \lambda^{n-s} \sum_{(s)} G^{(s)} F^{(s)} = 0 \tag{1.7-1}$$

remembering that $G^{(s)}$ and $F^{(s)}$ are corresponding s-rowed minors of the determinants of the G and F matrices and the first sum is over all possible such s-rowed minors (s). Such an equation is separately applicable to each symmetry block of the molecule. The number of different vibrational frequencies in the symmetry block is n in Eq. (1.7-1), counting degenerate frequencies only once.

Thus, for a symmetry block containing four frequencies, $n = 4$, and we have a quartic equation in λ. The first term, with $s = 0$, is then λ^4. The second term is $-\lambda^3[G_{11}F_{11} + G_{12}F_{12} + \cdots]$. The third term, with $s = 2$, is

$$\lambda^2 \left[\begin{vmatrix} G_{11} G_{12} \\ G_{21} G_{22} \end{vmatrix} \begin{vmatrix} F_{11} F_{12} \\ F_{21} F_{22} \end{vmatrix} + \begin{vmatrix} G_{11} G_{13} \\ G_{21} G_{23} \end{vmatrix} \begin{vmatrix} F_{11} F_{13} \\ F_{21} F_{23} \end{vmatrix} + \cdots \right].$$

The fourth term, with $s = 3$, is

$$-\lambda \left[\begin{vmatrix} G_{11} G_{12} G_{13} \\ G_{21} G_{22} G_{23} \\ G_{31} G_{32} G_{33} \end{vmatrix} \begin{vmatrix} F_{11} F_{12} F_{13} \\ F_{21} F_{22} F_{23} \\ F_{31} F_{32} F_{33} \end{vmatrix} + \cdots \right].$$

The term for $s = 4$ is just the determinant of the G-matrix times the determinant of the F matrix. Thus, we have the quartic equation

$$\lambda^4 - C_3\lambda^3 + C_2\lambda^2 - C_1\lambda + C_0 = 0 \qquad (1.7\text{-}2)$$

The coefficients C_i are functions of G_{ij} and F_{ij}. The G_{ij} can be readily calculated as described in Sect. 1.6. If we know the appropriate values for the F_{ij}, we can then calculate solutions for the four frequencies. Unfortunately, generally one wishes to go the other way: that is, one observes the four frequencies and wishes to calculate the force constants from them. A general quadratic valence force field is written as

$$2V = \sum_{ij} F_{ij}S_iS_j \qquad (1.7\text{-}3)$$

The S_i and S_j are internal coordinates and F_{ij} are force constants. For a given symmetry block, γ, we can write

$$2V^\gamma = \sum_{ij} F_{ij}^\gamma S_i^\gamma S_j^\gamma \qquad (1.7\text{-}4)$$

The S_i^γ are the symmetry coordinates while the F_{ij}^γ are the symmetry force constants. For a symmetry block of four frequencies there are ten force constants (including the generally smaller off-diagonal constants, F_{ij}^γ with $i \neq j$). Thus the only possibility for solving Eq. (1.7-2), with only four frequencies available, is to fix six of these constants and calculate the other four to reproduce the known frequencies. After fixing six of the constants, approximate values are chosen for the other four and adjusted by an iterative procedure to fit the observed frequencies. Dr. J. H. Schachtschneider (*13*) of the Shell Development Company has developed some excellent programs for calculating G matrices and for solving the secular equation using digital computers. Other information such as frequencies of isotopic species, centrifugal distortion constants, etc., will decrease the number of force constants which must be fixed (see Sect. 1.11).

Another complication arises because for a polynomial of degree n there are $n!$ solutions. Thus, for a 4×4 symmetry block, even after six force constants are fixed, there are 24 sets of the other four force constants which will reproduce the observed frequencies. Of course many of these solutions are imaginary and most of them will be unreasonable by comparison with other systems. However, there are often two or more solutions which are reasonable and thus further information, such as isotopic shifts, is required to remove the ambiguity. This will be discussed further in Chapter 2.

1.8. Vibrational Amplitudes and Potential Energy Distribution

In the solution of the secular equation (Sects. 1.6 and 1.7), the normal coordinates can be related to the symmetry coordinates. Thus, if we consider a symmetry block of n normal vibrations, we have the relations

$$\begin{aligned} S_1^\gamma &= L_{11}Q_1 + L_{12}Q_2 + \cdots L_{1n}Q_n \\ S_2^\gamma &= L_{21}Q_1 + L_{22}Q_2 + \cdots L_{2n}Q_n \\ &\ \ \vdots \\ S_n^\gamma &= L_{n1}Q_1 + L_{n2}Q_2 + \cdots L_{nn}Q_n \end{aligned} \tag{1.8-1}$$

For a normal mode of vibration in which the normal coordinate Q_k changes with frequency ν_k, all the symmetry coordinates involved change with the same frequency. However, for each symmetry coordinate, the amplitude is different. The relative amplitudes of the various symmetry coordinates are

$$(S_1^\gamma)_k/(S_2^\gamma)_k/(S_3^\gamma)_k/\cdots = L_{1k}/L_{2k}/L_{3k}/\cdots \tag{1.8-2}$$

The relative amplitudes L_{ij} can be evaluated from the relations

$$\begin{aligned} &(\textstyle\sum G_{1t}F_{t1} - \lambda)L_{1k} + (\sum G_{1t}F_{t2})L_{2k} + \cdots (\sum G_{1t}F_{tn})L_{nk} = 0 \\ &(\textstyle\sum G_{2t}F_{t1})L_{1k} + (\sum G_{2t}F_{t2} - \lambda_k)L_{2k} + \cdots (\sum G_{2t}F_{tn})L_{nk} = 0 \\ &\qquad\qquad\qquad\vdots \\ &(\textstyle\sum G_{nt}F_{t1})L_{1k} + (\sum G_{nt}F_{t2})L_{2k} + \cdots (\sum G_{nt}F_{tn} - \lambda_k)L_{nk} = 0 \end{aligned} \tag{1.8-3}$$

As usual $\lambda_k = 4\pi^2\nu_k^2$. Equation (1.8-3) is the expanded form of matrix Eq. (1.6-1). If we know the G values (from masses and geometry), the λ_k (from observed frequencies), and the F_{ij} (calculated as described in the previous section), we can calculate the relative amplitudes of the various symmetry coordinates for a given λ_k. The procedure is to evaluate the coefficients in (1.8-3) and solve for the various L_{ik} in terms of one of the L_{ik}, such as L_{1k}. This is completely analogous to the evaluation of relative amplitudes of motion for the atoms in linear ABC, treated in Appendix III.

It will often turn out that one of the L_{ik} is relatively large compared to the others in which case we ascribe the vibration to that particular symmetry coordinate. However, some chemists argue that a more specific and meaningful means of classifying a vibration as belonging to a particular type of coordinate is to calculate the potential energy distribution for a given normal vibration (*14*). Substitution of Eq. (1.8-1) into (1.7-4) for a given symmetry block, γ, gives

$$2V^{\gamma} = \sum_{n} \sum_{i,j} F_{ij}^{\gamma} L_{in} L_{jn} Q_n^2$$

for cross terms as $Q_k Q_l$ vanish.

For a given normal vibration

$$2V_k^{\gamma} = Q_k^2 \sum_{i,j} F_{ij}^{\gamma} L_{ik} L_{jk} \tag{1.8-4}$$

The various terms in the summation in (1.8-4) represent the potential energy of the various coordinates, assigning terms with $i \neq j$ to *coordinate interactions*. Generally the terms with $i = j$ are much larger and for a given normal vibration, k, one particular term, $F_{ll}^{\gamma} L_{lk}^2$, is the largest. We then assign that frequency to coordinate l. If there are two large terms of similar magnitude in the potential energy distribution, we say that the vibration is a strong mixture of the two coordinates involved. Generally the potential energy distribution is normalized so that

$$\sum F_{ij}^{\gamma} L_{ik} L_{jk} = 1$$

The amplitudes (eigenvectors) and potential energy distribution are included as output in the excellent computer programs developed by Schachtschneider (*13*).

1.9. General Quadratic Valence Force Field (GQVFF)

In Sect. 1.1 it was shown that cubic and higher terms in the potential function have a significant effect on the observed frequency. Thus, for $H^{35}Cl$ the neglect of cubic and higher terms results in an error of 3.5% for ω_e. This would mean an error of 7% in the force constant. This molecule has a large amount of anharmonicity and for most molecules (without hydrogen stretching vibrations) the anharmonicity is much smaller. Thus, neglect of cubic terms should lead to a fair approximation. Often it is possible to estimate the anharmonic corrections and extrapolate the observed frequencies to the ω_e (for example, see Sect. 1.3.1 and Appendix I). For the *frequencies* of *infinitesimal vibrations* $(\omega_e)_i$, it is a good approximation to neglect cubic and higher terms in the displacements for calculation of quadratic force constants.

The potential energy can be expressed as a power series in the internal displacement coordinates

$$2V = V_0 + \sum f_i S_i + \sum F_{ij} S_i S_j + \sum F_{ijk} S_i S_j S_k + \cdots \quad (1.9\text{-}1)$$

The S_i are displacements of internal coordinates from their equilibrium values. By an *internal coordinate* we mean an interatomic distance or bond angle which serves to define the molecular configuration. In the potential energy we must include a sufficient number of S_i to completely define the atomic arrangement in the molecule. Sometimes, to take advantage of symmetry, or for other reasons, we may include more than enough S_i to define the molecular configuration; in that case we have a *redundancy* (all coordinates are not independent) and special considerations are necessary. This is discussed in Sect. 1.10.

The zero of potential energy is arbitrary so we say $V_0 = 0$, meaning that in the equilibrium configuration the potential energy is zero. Where no redundancies are involved, the $f_i = 0$. This is because at equilibrium $(\partial V/\partial S_i)_0 = 0$ if all coordinates are independent. For small displacements, as discussed above, cubic and higher terms are negligible. Thus we are left with a general quadratic potential function, Eq. (1.9-2).

$$2V = \sum_{ij} F_{ij} S_i S_j \quad (1.9\text{-}2)$$

For correlation with chemical bonding theories, the most meaningful potential function is a General Quadratic Valence Force Field, in which the S_i and S_j are displacements of bond distances and angles between bonds. Because of this fact, throughout most of this book, a General Quadratic Valence Force Field, GQVFF, will be assumed. It will be modified when necessary; however, any specialized quadratic force field can be obtained by imposing certain restrictions on the GQVFF.

If, for an unsymmetrical molecule, there are n vibrational modes there are $n(n + 1)/2$ independent force constants.* The presence of symmetry in the molecule reduces the number of independent force constants, for, obviously, if there are two symmetrically equivalent XY bonds in the molecule they will have the same force constant, F_{XY}, and interactions involving either XY bond with a particular third coordinate will often be the same.

When the molecule is reduced to its different symmetry blocks, each symmetry block of n^γ normal coordinates (counting only once if degenerate)

* If F_{ij} and F_{ji} were independent, there would obviously be $n^2 F_{kl}$ for n vibrational modes. However, F_{ij} must equal F_{ji} so we subtract $(n^2 - n)/2$ arriving at $n(n + 1)/2$.

contributes $n^{\gamma}(n^{\gamma}+1)/2$ force constants. There is no interaction between coordinates of different symmetry. The total number of force constants is then $\sum_{\gamma} n^{\gamma}(n^{\gamma}+1)/2$ while the number of vibrational frequencies is $\sum^{\gamma} n_{\gamma}$. Therefore, unless all $n^{\gamma}=1$ as for linear, symmetrical YXY, there are more force constants than vibrational frequencies. The excess of force constants over frequencies equals $\sum_{\gamma} n^{\gamma}(n^{\gamma}-1)/2$.

As an example, consider square planar $M(CN)_4^{-x}$. The number of normal coordinates is broken up into the following representations: $2A_{1g}$, $1A_{2g}$, $2B_{1g}$, $2B_{2g}$, $1E_g$, $2A_{2u}$, $2B_{2u}$, and $4E_u$. The number n^{γ} preceding the symbol for the representation gives the number of normal coordinates in that representation, counting degenerate coordinates only once. Then the number of vibrational frequencies is $\sum n^{\gamma}=16$. The number of independent force constants is $\sum_{\gamma} n^{\gamma}(n^{\gamma}+1)/2=27$. Thus there are eleven more symmetry force constants than there are vibrational frequencies. In order to solve for the force constants, one must have additional information. Such information as may be available is described in a later section (1.11).

1.10. The Meaning of Force Constants of Polyatomic Molecules in a General Quadratic Force Field

1.10.1. Nonredundant Coordinates

The potential function for a General Quadratic Force Field with no redundancies, as shown in the previous section, is given by

$$2V=\sum_{i}^{n}\sum_{j}^{n} F_{ij}S_iS_j \tag{1.10-1}$$

The F_{ii} are the primary force constants; the F_{ij} are secondary, or interaction constants. S_i S_j, etc., are *displacements of internal coordinates* i, j, etc. For the GQVFF (General Quadratic Valence Force Field), i, j, etc., are bond lengths or angles included by bonds. We have stated there are no redundancies, meaning all coordinates in Eq. (1.10-1) are independent. A given F_{kk}, which is equal to $(\partial^2 V/\partial S_k^2)_0$, is the restoring force in coordinate k caused by a small unit displacement of coordinate k, keeping the other coordinates in Eq. (1.10-1) fixed. If k is a bond length, then F_{kk} is a measure of the bond strength* as for diatomic molecules (see Sects. 1.1 and 1.2). If k is a bond angle, then F_{kk} is a measure of the directional forces

* However, it is not completely comparable to a diatomic molecule force constant as it includes effects arising from changes in other internal coordinates.

maintaining the equilibrium configuration of that angle.* The meaning of the interaction constants is more vague. $F_{ij} = (\partial^2 V/\partial S_i \, \partial S_j)_0$ mathematically; however, its physical meaning is somewhat obscure. We shall now try to clarify it. Suppose we impose a displacement of $+1$ for k ($S_k = +1$). Then, from (1.10-1)

$$2V = F_{kk} + \sum_{i \neq k}^{n} \left(2F_{ik}S_i + \sum_{l \neq k}^{n} F_{il}S_iS_l \right) \tag{1.10-2}$$

Thus, if we force S_k to be $+1$ and keep all other S_i equal to zero, the potential energy increases by $(\frac{1}{2})F_{kk}$. However, if we constrain coordinate k to a displaced position, the electronic configuration will change and tend to cause a displacement of other internal coordinates so as to minimize the potential energy. At this new minimum, we write from (1.10-2)

$$\frac{\partial V}{\partial S_l} = 0 = F_{lk} + \sum_{i \neq k}^{n} F_{il}(i)_k, \qquad l = 1, n; \quad l \neq k \tag{1.10-3}$$

The term $(i)_k$ is short for the displacement of coordinate i which would minimize the energy after unit positive displacement of coordinate k. We can then write a rather unwieldy definition of F_{kl}

$$F_{kl} = -\sum_{i \neq k}^{n} F_{il}(i)_k, \qquad l = 1, n; \quad l \neq k \tag{1.10-4}$$

Since the F_{il} are generally small compared to F_{ll} and the $(i)_k$ can be expected to be of the same order of magnitude as $(l)_k$,

$$F_{kl} \approx -F_{ll}(l)_k \tag{1.10-5}$$

This serves as an approximate definition of F_{kl}. Thus *the force constant for interaction of coordinate k with coordinate l, F_{kl}, is approximately proportional to the displacement of coordinate l resulting from minimization of energy after unit displacement of coordinate k.*†

This approximation (1.10-5) is good only for relatively large values of F_{kl} and F_{ll}. Generally to determine the $(l)_k$, which we call "interaction coordinates," we must solve the set of simultaneous equations (1.10-4).

* However, it is important to realize that the value of F_{kk} depends on what other coordinates are used to define the potential field in (1.10-1). This is discussed further in Sect. 1.13.

† Another definition for F_{kl}, as sometimes given, is the force imposed on coordinate k resulting from unit displacement of coordinate l. This is a less informative measure of interaction as discussed in Appendix IV.

The solution, by Cramer's Rule, is

$$(l)_k = \{F_{lk}\}/\{F_{kk}\} \tag{1.10-6}$$

where $\{F_{ij}\}$ is the ij cofactor of the F matrix. The cofactor $\{F_{ij}\}$ is $(-1)^{i+j}\,|M_{ij}^F|$ where $|M_{ij}^F|$ is the determinant of the i, jth minor of the F matrix.

The set of equations (1.10-4) can be arrived at in a more mathematically satisfying way. The potential energy is given as above, by Eq. (1.10-1). Let us substitute dR_i for S_i, etc. We wish to obtain (dR_i/dR_k) with the constraint that the potential energy remain a minimum. Therefore we set $(\partial V/\partial R_l) = 0$ for all l except k, arriving at the set of equations

$$\sum_{i=1}^{n} F_{il}\, dR_i = 0 \qquad \text{with} \quad l = 1, n; \quad l \neq k$$

Differentiating each equation with respect to R_k we arrive at Eq. (1.10.3) with $(i)_k = (dR_i/dR_k)_{V=\min}$.

The interaction coordinates are calculated much more readily from compliance constants. The compliance constant matrix is the inverse of the force constant matrix ($C = F^{-1}$). Therefore, the elements of C are related to F by, $C_{ij} = \{F_{ij}\}/|F|$, where again $\{F_{ij}\}$ is the cofactor of F, and $|F|$ is the determinant of F. Then it is apparent that Eq. (1.10-6) reduces to

$$(l)_k = C_{lk}/C_{kk} \tag{1.10-7}$$

Compliance constants are discussed further in Sect. 1.13.

Interaction coordinates, $(l)_k$, are rather interesting quantities from a chemist's point of view. One can hope to calculate from molecular orbital theory what the minimum energy configuration will be if one coordinate is constrained in a configuration displaced from its normal equilibrium value. Such a calculation would give the interaction coordinate, and thus would be an important link of vibrational spectra to electronic structure. It is possible that such a calculation will not be very sensitive, but even so the $(l)_k$ give one at least a qualitative picture of changes in electronic configuration accompanying a constrained displacement of one coordinate.

1.10.2. Redundant Coordinates

For some systems, a redundancy exists among some of the coordinates; that is, they are not independent. A change in one of these coordinates cannot take place without a change in one or more other coordinates. For example, consider the six bond angles of tetrahedral XY_4 (as in CH_4). If one angle is increased, there must take place decreases in one or more of the other five bond angles. In general, the redundancy can be written as

$$\sum_{i} a_i S_i^{\mathrm{R}} + \sum_{ij} b_{ij} S_i^{\mathrm{R}} S_j^{\mathrm{R}} + \sum_{ijk} c_{ijk} S_i^{\mathrm{R}} S_j^{\mathrm{R}} S_k^{\mathrm{R}} + \cdots = 0$$

The superscript R refers to the redundant group of coordinates. For a quadratic potential function, we are not concerned with cubic and higher terms in the displacement coordinates, S_1.

If such a redundancy exists we can no longer say $(\partial V/\partial S_k^R)_0 = 0$; instead when S_k^R is changed there must be a change in some other S_i^R and $(\partial V/\partial S_k^R)_0 = \rho a_k$, with ρ an undetermined multiplier.* Thus we must have a term first order in the S_i^R in the potential function

$$2V = \rho \sum_{l} a_l S_l^{\mathrm{R}} + \sum_{ij} F_{ij} S_i S_j \tag{1.10-8}$$

The second summation includes redundant and nonredundant coordinates.

The redundancy and the presence of the linear term in (1.10-8) lead to some obscurity in the meaning of the force constants involved. For a while, due to an error in matrix manipulation, it was thought that the symmetry force constant for a redundant coordinate was zero. Thus, for tetrahedral XY_4, the redundant coordinate (to the first order) of A_1 symmetry is $\sum_{i>j} \alpha_{ij} = 0$, where α_{ij} is the H_iCH_j angle. According to Gold et al. (*15*), this redundancy leads to the conclusion that the corresponding A_1 symmetry force constant, $F_\alpha + 4F'_{\alpha\alpha} + F''_{\alpha\alpha} = 0$. F_α is the primary H_iCH_j angle bending force constant. $F'_{\alpha\alpha}$ is the constant for interaction of α_{ij} with α_{ik} (adjacent angles) and $F''_{\alpha\alpha}$ is the constant representing interactions of α_{ij} with α_{kl} (opposite angles). However, Crawford and Overend (*16*) have refuted the argument (*15*) and shown that $F_\alpha + 4F'_{\alpha\alpha} + F''_{\alpha}$ is completely indeterminate for XY_4, and, in general, symmetry force constants for redundant symmetry coordinates are indeterminate.

With this background, we can look into the meaning of force constant involving redundant coordinates. As the classic example, we shall consider tetrahedral XY_4, for which the General Quadratic Potential function can be written

$$\begin{aligned} 2V = {} & 2\rho \sum_{i<j} \alpha_{ij} + F_r \sum_{i} r_i^2 + F_\alpha \sum_{i<j} \alpha_{ij}^2 + 2F'_{rr} \sum_{i<j} r_i r_j + 2F'_{\alpha\alpha} \sum_{\substack{j<k \\ i \neq j,k}} \alpha_{ij}\alpha_{ik} \\ & + 2F''_{\alpha\alpha} \sum_{\substack{i<j,k \\ k<l \\ j \neq k}} \alpha_{ij}\alpha_{kl} + 2F'_{r\alpha} \sum_{i \neq j} r_i \alpha_{ij} + 2F''_{r\alpha} \sum_{\substack{j<k \\ k,j \neq i}} r_i \alpha_{jk} \end{aligned} \tag{1.10-9}$$

* This has been discussed in detail in (6), p. 172.

All r_i and α_{ij} in (1.10-9) are small displacements, really representing Δr_i and $\Delta\alpha_{ij}$. The α_{ij} are redundant and the second order relation is

$$\sum_{i<j}\alpha_{ij} + \left(\frac{3}{4\sqrt{2}}\right)\sum_{i<j}\alpha_{ij}^2 + \left(\frac{1}{\sqrt{2}}\right)\sum_{\substack{j<k\\ i\neq j,k}}\alpha_{ij}\alpha_{ik} = 0 \tag{1.10-10}$$

The first-order term can now be removed by substituting the second-order expression for $\sum_{i<j}\alpha_{ij}$ from Eq. (1.10-10) into (1.10-9). We find

$$2V = \left(F_\alpha - \frac{3}{2\sqrt{2}}\rho\right)\sum_{i<j}\alpha_{ij}^2 + 2\left(F'_{\alpha\alpha} - \frac{1}{\sqrt{2}}\rho\right)\sum_{\substack{j<k\\ i\neq j,k}}\alpha_{ij}\alpha_{ik} + 2F''_{\alpha\alpha}\sum_{\substack{i<j,k\\ k<l\\ j\neq k}}\alpha_{ij}\alpha_{kl} + \cdots \tag{1.10-11}$$

The other quadratic terms are the same as in (1.10-9).

There is a further simplification to make. In Eq. (1.10-9) [and (1.10-11)] due to the first-order redundancy of (1.10-10), $\sum_{j<k,k,j\neq i} r_i\alpha_{jk} = -\sum_{i\neq j} r_i\alpha_{ij}$. For example, consider the term $r_1(\alpha_{23} + \alpha_{24} + \alpha_{34})$. From the first-order redundancy [first term of (1.10-8)], $\alpha_{23} + \alpha_{24} + \alpha_{34} = -\alpha_{12} - \alpha_{13} - \alpha_{14}$. The second order term is not included because when multiplied by r_i (which is short for Δr_i), it contributes only in third order. Therefore, the last two terms in (1.10-9) can be written as $2(F'_{r\alpha} - F''_{r\alpha})\sum_{i\neq j} r_i\alpha_{ij}$. This shows that $F'_{r\alpha}$ and $F''_{r\alpha}$ occur only as their difference and cannot be determined independently. In fact, by themselves $F'_{r\alpha}$ and $F''_{r\alpha}$ have no meaning, so it is pointless to carry them as separate constants. Therefore, we shall say $F'_{r\alpha} - F''_{r\alpha} = h_{r\alpha}$ and try to define the latter quantity. We shall also define $h_\alpha = F_\alpha - (3/2\sqrt{2})\rho$ and $h'_{\alpha\alpha} = F'_{\alpha\alpha} - (1/\sqrt{2})\rho$, then

$$2V = F_r\sum_i r_i^2 + 2F_{rr}\sum_{i<j} r_i r_j + h_\alpha\sum_{i<j}\alpha_{ij}^2 + 2h'_{\alpha\alpha}\sum_{\substack{j<k\\ i\neq j,k}}\alpha_{ij}\alpha_{ik} + 2F''_{\alpha\alpha}\sum_{\substack{i<j,k\\ k<l\\ j\neq k}}\alpha_{ij}\alpha_{kl} + 2h_{r\alpha}\sum_{i\neq j} r_i\alpha_{ij} \tag{1.10-12}$$

Now to define the force constants, it is convenient to remove the redundancy by omitting one of the α_{ij} from (1.10-12). From (1.10-10) to the first order

$$\alpha_{34} = -\alpha_{12} - \alpha_{13} - \alpha_{14} - \alpha_{23} - \alpha_{24} \tag{1.10-13}$$

remembering that α_{ij} is a small displacement of the Y_iXY_j angle. We do not worry about the second-order term as all the displacements in (1.10-12) are quadratic so the second-order terms in α_{ij} contribute to third or fourth order. Substituting (1.10-13) into (1.10-12) leads to

$$
\begin{aligned}
2V = {} & F_r \sum_i r_i^2 + 2F_{rr} \sum_{i<j} r_i r_j + 2(h_\alpha - h'_{\alpha\alpha})(\alpha_{13}^2 + \alpha_{14}^2 + \alpha_{23}^2 \\
& + \alpha_{24}^2 + \alpha_{13}\alpha_{14} + \alpha_{13}\alpha_{23} + \alpha_{14}\alpha_{24} + \alpha_{23}\alpha_{24} + 2\alpha_{13}\alpha_{24} + 2\alpha_{14}\alpha_{23}), \\
& + 2(h_\alpha - F''_{\alpha\alpha})(\alpha_{12}^2 + \alpha_{12}\alpha_{13} + \alpha_{12}\alpha_{14} + \alpha_{12}\alpha_{23} + \alpha_{12}\alpha_{24} - \alpha_{13}\alpha_{24} \\
& - \alpha_{14}\alpha_{23}) + 2h_{r\alpha}[r_1(\alpha_{12} + \alpha_{13} + \alpha_{14}) + r_2(\alpha_{12} + \alpha_{23} + \alpha_{24}) \\
& - r_3(\alpha_{12} + \alpha_{14} + \alpha_{24}) - r_4(\alpha_{12} + \alpha_{13} + \alpha_{23})]
\end{aligned} \tag{1.10-14}
$$

In this equation all the displacements are mathematically independent; the redundancy is removed. It is apparent that h_α, $h'_{\alpha\alpha}$, and $F''_{\alpha\alpha}$ cannot be determined independently. All we can hope to determine is $h_\alpha - h'_{\alpha\alpha}$ and $h_\alpha - F''_{\alpha\alpha}$ or linear combinations of these differences.

Inasmuch as the displacements in (1.10-14) are independent, we can investigate the meaning of the force constants, as was done for the general case using Eqs. (1.10-2–1.10-4).

F_r of course is the restoring force of an XY bond after it has suffered a small displacement. It is a measure of the strength of the XY bond although, as mentioned previously, it is not unique but depends on the assumed force field.

To define F_{rr}, first consider a forced displacement of $r_1 = +1$, thus

$$
2V = F_r + F_r \sum_{i \neq 1} r_i^2 + 2F_{rr}(r_2 + r_3 + r_4) + 2F_{rr} \sum_{\substack{i \neq 1 \\ i<j}} r_i r_j + \cdots
$$

Then take

$$
\begin{aligned}
\partial V / \partial r_2 = 0 = {} & F_r(r_2)_1 + F_{rr} + F_{rr}[(r_3)_1 + (r_4)_1] \\
& + h_{r\alpha}[(\alpha_{12})_1 + (\alpha_{23})_1 + (\alpha_{24})_1]
\end{aligned}
$$

As before $(r_i)_k$ and $(\alpha_{ij})_k$ are displacements of r_i and α_{ij} which would lead to minimization of energy after unit small displacement of coordinate k. By symmetry $(r_2)_1 = (r_3)_1 = (r_4)_1$ and $(\alpha_{23})_1 = (\alpha_{24})_1 = -(\alpha_{12})_1$; then

$$
F_{rr} = \frac{-F_r(r_2)_1 + h_{r\alpha}[(\alpha_{12})_1]}{1 + 2(r_2)_1} \tag{1.10-15}
$$

Similar arguments lead to the following relations:

$$
h_{r\alpha} = \frac{-2(h_\alpha - F''_{\alpha\alpha})(\alpha_{12})_1}{1 - (r_2)_1} \tag{1.10-16}
$$

$$
h_{r\alpha} = \frac{-(F_r - F_{rr})(r_1)_{12}}{1 + 2(\alpha_{13})_{12}} \tag{1.10-17}
$$

$$
h_\alpha - F''_{\alpha\alpha} = \frac{-6(h_\alpha - h'_{\alpha\alpha})(\alpha_{13})_{12} - 2h_{r\alpha}(r_1)_{12}}{1 - (\alpha_{13})_{12}} \tag{1.10-18}
$$

Equations (1.10-15–1.10-18) serve as physical definitions of the constants F_{rr} and $h_{r\alpha}$. As mentioned previously h_α, $h'_{\alpha\alpha}$, and $F''_{\alpha\alpha}$ cannot be defined separately; thus, we can only write relations among them, such as (1.10-18). In these equations it must be remembered that $(r_2)_1$ is the change in r_2 leading to minimization of energy after small unit change in r_1; $(r_1)_{12}$ is the change in r_1 leading to minimization of energy after small unit change of α_{12}, etc.

From Eq. (1.10-14) we might define $2(h_\alpha - h'_{\alpha\alpha})$ as the restoring force, resulting from unit displacement of α_{13} holding α_{14}, α_{23}, α_{24}, and α_{12} fixed, and $2(h_\alpha - F''_{\alpha\alpha})$ as the restoring force resulting from unit displacement of α_{12} holding α_{13}, α_{14}, a_{23}, and α_{24} fixed. This can be expressed mathematically as

$$2(h_\alpha - h'_{\alpha\alpha}) = (\partial^2 V/\partial\alpha_{13}^2)_{\alpha_{12},\alpha_{14},\alpha_{23},\alpha_{24}} = 0, \qquad \alpha_{34} = -\alpha_{13}$$
$$2(h_\alpha - F''_{\alpha\alpha}) = (\partial^2 V/\partial\alpha_{12}^2)_{\alpha_{13},\alpha_{14},\alpha_{23},\alpha_{24}} = 0, \qquad \alpha_{34} = -\alpha_{12}$$

Looking at Eqs. (1.10-15) and (1.10-16), we see that if we determine F_{rr}, $h_{r\alpha}$, and $(h_\alpha - F''_\alpha)$ from the vibrational spectra and supplementary data, we can determine what changes in r_2 and α_{12} would lead to minimum energy if r_1 is constrained to a small positive unit displacement. These "interaction displacement coordinates" are, of course, directly related to changes in electronic structure accompanying such a forced displacement of r_1 and are thus important quantities for correlation with bonding theory.

A similar discussion has been given by the author for square planar XY_4 (*17*) and for planar XY_3 (see Appendix VI of this book). In Sect. 1.13 it will be shown that the use of compliance constants leads to much simpler and more straightforward relations.

1.11. Supplementary Data for Force Constant Calculation

It has been pointed out that for all but the simplest molecules, the number of vibrational frequencies is insufficient for determination of all the force constants in a General Quadratic Force Field. There are other sources of information which we shall discuss below. In this discussion, a General Quadratic Force Field is assumed.

1.11.1 Vibrational Frequencies of Isotopic Species

Since isotopic molecules have essentially the same electronic structure, the potential function under the influence of which the nuclei move relative to each other should be the same to a high order of approximation. Therefore, we can expect that force constants are invariant to isotopic substitution. In Sect. 2.3 of this book, it is shown that this is indeed a good approximation. We shall now proceed with the knowledge that force constants are invariant to isotopic substitution to a high degree of accuracy.

The question arises: "How much additional information can we obtain from the frequencies of isotopic species?" It is clear that for a diatomic molecule, such as $^1H^{35}Cl$, the fundamental harmonic frequency is sufficient information to calculate the force constant (Eq. 1.1-3). A knowledge of the fundamental frequency of $^2H^{35}Cl$ or $^1H^{37}Cl$ also allows calculation of the HCl force constant, serving as a check on the calculation for $^1H^{35}Cl$ but giving no new information. However, for an unsymmetrical linear triatomic molecule this is not the case. In Appendix III, we see that the bending force constant is determined by the bending frequency of one molecular species. However, from Eq. (III-6), we see that for the two stretching vibrations of linear ABC,

$$\lambda^2 - C_1\lambda + C_0 = 0$$
$$C_1 = F_{AB}(\mu_A + \mu_B) + F_{BC}(\mu_B + \mu_C) - 2F_{AB,BC}\mu_B$$
$$C_0 = (F_{AB}F_{BC} - F^2_{AB,BC})(\mu_A\mu_B + \mu_B\mu_C + \mu_A\mu_C)$$
$$\lambda = 4\pi^2\nu^2, \qquad \mu_k = 1/M_k$$

From the properties of a polynomial equation we know that $C_1 = \lambda_1 + \lambda_2$ and $C_0 = \lambda_1\lambda_2$. There are only two frequencies, but three force constants, F_{AB}, F_{BC}, and $F_{AB,BC}$. The stretching frequencies for the isotopic molecule, ^iABC, give us two additional pieces of information. However, the product ratio $\lambda_1\lambda_2/\lambda_1^i\lambda_2^i = C_0/C_0^i = (\mu_A\mu_B + \mu_B\mu_C + \mu_A\mu_C)/(\mu_A^i\mu_B + \mu_B\mu_C + \mu_A^i\mu_C)$ does *not depend on the force constants.* This is another way of stating the Teller–Redlich product rule (7). Thus, if we know λ_1, λ_2, and λ_1^i, then λ_2^i is fixed by the product ratio defined by the masses and geometry, independent of force constants. Therefore, the four stretching frequencies of ABC and ^iABC result in three independent pieces of information for force constant calculation. This is sufficient as there are only three stretching force constants. Subsequent substitution of ^jB for B would give us no additional information except as a check on the calculations. In other words, after sufficient isotopic species have been studied to calculate the general quadratic force constants, further isotopic substitution merely gives improved statistics on the results. Of course, this additional data is generally useful as it enables one to determine statistical error limits. For the case of large mass changes, such as D for H, this additional data may be useful for determination of anharmonicity corrections (*19*).

As a slightly more complicated case, consider the unsymmetrical bent triatomic molecule, XYZ. For such a molecule the stretching and bending coordinates can mix in the quadratic approximation and there are six force constants related to three vibrational frequencies. One isotopic substitution, X^iYZ, gives us three new frequencies but again the product ratio, $\lambda_1\lambda_2\lambda_3/\lambda_1^i\lambda_2^i\lambda_3^i = |G|/|G^i|$, is independent of force constants.* Thus, from the two isotopic species there are only five independent frequencies for force constant calculation. A third isotopic species, such as XY^iZ, is necessary. This would give two more data, which is more than sufficient.

Symmetry, of course, reduces the number of different force constants and thus reduces the number of isotopic species required. The following remarks apply to complete isotopic substitution of a symmetrically equivalent set of atoms. The number of frequencies in a given symmetry block will determine the number of isotopic species necessary for evaluation of the symmetry force constants in that symmetry block. In the general case, if there are n genuine vibrations in a given symmetry block (or $N = (n/3) + 2$ atoms in a completely unsymmetrical molecule), each isotopic substitution will give no more than $n - 1$ additional independent pieces of information useful for force constant calculations, because the product ratio fixes one of the frequencies relative to the others. Since there are $n(n + 1)/2$ general quadratic force constants for a symmetry block of n frequencies, it is required to have at least $n/2$ isotopic substitutions, that is, vibrational frequencies from a total of $(n/2) + 1$ isotopic molecules. Thus, a symmetry block of six vibrations (or an unsymmetrical molecule of four atoms) requires at least four isotopic molecules to determine the general quadratic force constants. As mentioned above, this discussion applies to complete isotopic substitution of a symmetrically equivalent set. Thus, for $Ni(CO)_4$ it applies for substitution of ^{13}C for all carbons, or ^{18}O for all oxygens. Partial isotopic substitution, such as $Ni(^{13}CO)(^{12}CO)_3$, will give no new force constant information which cannot be obtained from $Ni(^{12}CO)_4$ and $Ni(^{13}CO)_4$. If complete substitution is not possible, one can often obtain similar information from partially substituted species; however, the symmetry is reduced and larger secular equations must be handled. Also the anharmonicity corrections bear no simple relationship among isotopic species of different symmetry.

* $|G|$ is the determinant of the G matrix (see Sect. 1.6).

The above discussion concerns the minimum number of isotopic species required. In many cases the vibrational frequencies of more isotopic molecules may be necessary because of additional relations among the frequencies or lack of sensitivity to a given mass change. Each case must be considered under its own merit. One of the disadvantages of use of isotopic substitution for force constant calculations is the lack of sensitivity of the isotopic shifts to the force constants. Therefore, it is important to have very accurate frequency data (better than $\pm 0.5\ cm^{-1}$ for isotope shifts) to obtain meaningful results. Also the anharmonicity corrections must be taken into account if at all possible. Without such corrections, large errors may enter for very anharmonic frequencies such as hydrogen stretching vibrations.

A word should be said about "double isotopic substitution" of a given atom. By "double isotopic substitution," the author means substitution first by one isotope, then by another isotope for a given atom or symmetrically equivalent set of atoms. In some cases, the second substitution gives no new information; in others it does give new information which, however, is very insensitive to the force constants. For example, in the molecule ONF, which is bent, substitution of ^{16}O by ^{18}O gives three new vibrational frequencies and two new pieces of information for force constant calculation. A second substitution, with ^{17}O, gives three more vibrational frequencies but only one new relation among force constants. Furthermore, this relation is too insensitive to be useful in determining the force constants though it would be useful as additional statistical data. The computer programs developed by J. H. Schachtschneider of The Shell Development Company are excellent for least squares estimation of force constants from vibrational frequencies of isotopic species.

1.11.2. Mean Square Amplitudes, Centrifugal Distortion Constants, and Coriolis Coupling Constants

Mean square amplitudes as determined from electron diffraction (*20*), centrifugal distortion constants (*21–24*, *33*) from infrared and microwave studies, and Coriolis coupling constants (*25*, *25a*, *26*) can theoretically supply additional information on force constants. These all apply to gas phase studies. Perhaps some day it will be possible to determine mean square amplitudes from X ray studies of solids to gain additional information on force constants. At present the results in the literature on mean square amplitude-force constant studies are not too encouraging. For example, Iwasaki and Hedberg (*27*) calculated the potential constants for

PCl_3 by combining mean square amplitudes from electron diffraction with vibrational frequencies. Mirri and co-workers (*28*) also calculated the potential constants for PCl_3 by combining vibrational frequency data with centrifugal distortion constants determined from microwave studies. The two sets of results are given for comparison in Table 1.11-1. The electron diffraction results leave rather large limits of error but as the method is further developed these should be improved. However, rather discouraging is the lack of agreement of the two columns of Table 1.11-1. Thus some of the constants of Mirri et al. (*28*) particularly f_r, do not fall in the very generous range of values given by Iwasaki and Hedberg (*27*).

TABLE 1.11-1

POTENTIAL CONSTANTS[a] OF PCl_3 FOR A GENERAL FORCE FIELD

	Iwasaki and Hedberg (*27*)	Mirri et al. (*28*)
f_r	2.83 ± 0.37	2.047 ± 0.005
f_{rr}	0.59 ± 0.30	0.366 ± 0.048
f_α	0.28 ± 0.04	0.343 ± 0.006
$f_{\alpha\alpha}$	0.05 ± 0.04	0.009 ± 0.007
$f'_{r\alpha}$	0.07 ± 0.28	0.005 ± 0.002

[a] Values are expressed in mdyn $Å^{-1}$.

Further discussion of mean square amplitudes is included under individual molecules later in this book.

In a similar calculation, Pillai and Gnanadesikan (*29*) calculated force constants for NH_2CN and ND_2CN using vibrational frequencies and centrifugal distortion constants. They also calculated the mean square amplitudes and found that the C≡N stretching amplitude compared favourably with the experimental value in HCN and the N—H and N—D stretching amplitudes were "slightly higher" than the experimental values in NH_3 and ND_3. However, this tells us nothing about the sensitivity of amplitudes to force constants. It appears that centrifugal distortion constants are much more useful than mean amplitudes for force constant calculations. This will be discussed further for nitrosyl halides, Sect. 2.13.

Duncan and Mills (*26*) have used observed Coriolis coupling constants along with vibrational frequencies to determine the potential constants of some XH_4 and XH_3 molecules. They show that the data from D-substituted species, though theoretically sufficient, is not in practice sensitive enough for calculation of the general force field without use of the

Coriolis coupling constants. This is largely because of the uncertainty in anharmonicity corrections necessary. In general it appears that the Coriolis coupling constants are quite sensitive to force constants and, thus, should be quite helpful in determination of the latter when good values are available. Further discussion is given later in this book under individual molecules.

Thus, theoretically it is possible to determine all the potential constants of a General Quadratic Force Field for some of the simpler polyatomic molecules. However, at present all of the data supplementary to vibrational frequencies are limited to the gas phase except for isotopic substitution. For solid state studies, we are faced with a deficiency of data for calculation of potential constants. We, therefore, must resort to studies of isotopic species when possible. For the gas phase, centrifugal distortion constants and Coriolis coupling constants are quite sensitive to force constants and are, thus, extremely useful in their determination. Mean square amplitudes, however, are not particularly useful for force constant determination; on the other hand, mean square amplitudes as calculated from force constants may be quite useful for interpretation of electron diffraction results.

1.12. Force Fields other than the GQVFF

The previous discussion in this book has been limited to a General Quadratic Valence Force Field (GQVFF) for which the internal coordinates are distances between bonded atoms and angles included by bonds. Another equally general quadratic force field is the General Quadratic Central Force Field (GQCFF) for which all the internal coordinates are interatomic distances (bonded and nonbonded). Since they are both completely general, from a mathematical viewpoint it is arbitrary which we use; however, one may be much more meaningful than the other from a chemical bonding viewpoint. A specific example of such a comparison is given in the treatment of the water molecule to follow in Sect. 2.11 (as well as treatment of ONF in Sect. 2.13). For both of these molecules, it is shown that the GQVFF makes much more chemical sense than the GQCFF.

Urey and Bradley (*30*), suggested a somewhat less general force field for tetrahedral molecules, such as methane, which is really a combination of simple* valence forces and simple* central forces. Thus, for tetrahedral

* By "simple" is meant interactions between coordinates are neglected except as implied by redundancies among the internal coordinates used.

XY_4, Urey and Bradley assumed valence forces between X and Y, directional valence forces determining the YXY angles, and essentially van der Waals' forces between Y atoms. Schimanouchi (*31*) has developed this approach further and many authors have treated various polyatomic molecules using this Urey–Bradley Force Field (UBFF). Such a force field has the great advantage that there are generally fewer force constants than in a general quadratic field. However, Duncan (*32*) has shown that even for fairly simple molecules such as CH_4, GeH_4, CF_4, NH_3, etc., the Urey–Bradley Force Field is unsatisfactory. In order to achieve good results, consistent with Coriolis coupling constants, etc., it is necessary to add stretch–stretch interaction force constants giving what is referred to as a "Modified Urey–Bradley Force Field" (MUBFF). Including such interactions increases the number of parameters so as to make the Urey–Bradley force field less useful and in some cases invalid. Duncan's results (*32*) did suggst that for the planar boron trihalides, the UBFF is quite satisfactory. However, as is discussed in Chapter 3, there still remains considerable doubt concerning the experimental results for these molecules and so this agreement must be accepted with reservations. The Urey–Bradley Force Field for trigonal planar molecules is treated in Appendix VII as a useful example of the method. It should be pointed out that the Urey–Bradley force constants can always be transformed to General Quadratic Force Constants as they arise merely from special constraints on the latter.

1.13. Compliance Constants

Decius (*33*) has presented an excellent discussion of compliance constants. Much of it is repeated below. In the following discussion, X' is the transpose of matrix X, X^{-1} is the inverse, S is a column matrix of internal displacement coordinates, ΔR_i. The matrix of compliance constants is the inverse of the force constant matix

$$C = F^{-1} \tag{1.13-1}$$

In Sect. 1.6 it is shown that in matrix form the secular equation can be written as $GFL = L\Lambda$ or $FG(L^{-1})' = (L^{-1})'\Lambda$. By a little matrix manipulation, it can be shown that $F^{-1}G^{-1}L = L\Lambda^{-1}$ and $G^{-1}F^{-1}(L')^{-1} = (L')^{-1}\Lambda^{-1}$. Defining the K matrix as $K = G^{-1}$ and the φ matrix as $\varphi = \Lambda^{-1}$, we find

$$CKL = L\varphi \tag{1.13-2}$$

$$KC(L')^{-1} = (L')^{-1}\varphi \tag{1.13-3}$$

$\varphi = \Lambda^{-1}$ is the diagonal matrix of the terms $(1/4\pi^2\nu_i^2)$.

Equations (1.13-2) and (1.13-3) give the two secular determinants, $|CK - \varphi E| = 0$ and $|KC - \varphi E| = 0$, with eigenvectors L and $(L')^{-1}$, respectively.

If we define a generalized force vector as

$$\mathscr{F} = -FS \tag{1.13-4}$$

so

$$S = -F^{-1}\mathscr{F} = -C\mathscr{F} \tag{1.13-5}$$

or

$$S_i = dR_i = -\sum_j C_{ij}f_j, \qquad i = 1, n \tag{1.13-6}$$

we can write the potential energy as

$$V = \tfrac{1}{2}S'FS = \tfrac{1}{2}\mathscr{F}'C\mathscr{F} \tag{1.13-7}$$

or

$$2V = \sum_{ij} C_{ij}f_if_j \tag{1.13-8}$$

The f_i represent force vectors imposed on the coordinates, R_i; they are elements of the $\mathscr{F}$ matrix. Thus, if 0.01 mdyn of force is applied toward increasing the length of bond k, this bond will stretch by $(0.01\ C_{kk})$ Å and the potential energy will increase by $(10^{-4}\ C_{kk}/2)$ mdyn Å if all other f_i are zero; that is, if the potential energy is constrained to a minimum after R_k is increased. Therefore, a primary (or diagonal) compliance constant is a measure of the displacement which will take place in a coordinate as a result of a force imposed on this coordinate, if the other coordinates are allowed to adjust themselves to minimize the energy. The off-diagonal compliance constants are much easier to define in a physical sense than are the corresponding force constants. The compliance constants are quite simply related to the interaction coordinates. Thus, Eq. (1.10-7) shows us that $C_{lk} = (l)_kC_{kk}$ where the interaction coordinate $(l)_k$ is the relative displacement of coordinate l if the energy is minimized with k constrained in a displaced configuration.

Equation (1.10-7) follows readily from Eqs. (1.13-6) which define the compliance constants. If we apply a force to coordinate k, f_k, and keep the energy a minimum, all the other coordinates will displace to relieve any force vectors arising from a change in electronic configuration. From Eq. (1.13-6), $(dR_k) = -C_{kk}f_k$ and $(dR_l) = -C_{lk}f_k$. Then $dR_l/dR_k = (l)_k = C_{lk}/C_{kk}$ as in Eq. (1.10 7).

Decius (*33*) has shown that a compliance constant for a given coordinate does not depend on what other coordinates are used to define the system. This is not true of force constants; for example, the calculated force constant of a given bond stretch is dependent upon what other coordinates are used to define the potential energy. This is a distinct advantage in favor of describing the force field in terms of compliance constants. As an illustrative example of this advantage, the water molecule is discussed in detail in Sect. 2.11 of this book.

A program has been written by Ottinger (*34*) and added to by Ryan (*35*) for a least squares estimate of compliance constants from vibrational frequencies, centrifugal distortion constants, and mean square amplitudes as input data. The secular Eq. (1.13-3), $KC(L')^{-1} = (L')^{-1}\varphi$, is solved to fit the vibrational frequencies. One can evaluate the $K = G^{-1}$ matrix elements as outlined by Polo (*36*) to solve the problem. However, with the availability of computers, it may often be easier to start with the G matrix and invert it to arrive at the K matrix. If redundancies exist, the G matrix rows and columns associated with the redundancies must be eliminated (*33*) before the matrix is inverted (otherwise G is singular and cannot be inverted). Similarly the corresponding rows and columns in the C matrix are eliminated (*37*).

The author has recently (*37*) come to the conclusion, stated by Decius (*33*) some years ago, that compliance constants are more interesting than force constants for discussion of bonding and electronic structure in molecules. The reasons for this conclusion are outlined below.

(1) Compliance constants are well-defined, independent of the coordinate system chosen. This is not true of force constants. Therefore, compliance constants are certainly unique and in many cases more appropriate for transfer from one molecule to similar molecules. This is especially true for interaction constants.

(2) The compliance constants are simply related to the interaction coordinates. The latter quantities are closely related to electronic structure and changes in electronic configuration accompanying a coordinate displacement.

(3) Because of the simple relation of compliance constants to interaction coordinates, a better estimate of the variance–covariance matrix for interaction coordinates is possible from compliance constants than from force constants.

(4) Force constants for coordinates involved in redundancies are indeterminate; only linear combinations can be determined (see Sect. 1.10.2). Compliance constants, on the other hand, are *uniquely defined* (*33*), though there are relations among them if redundancies exist.

(5) Compliance constants come more naturally than force constants into the potential constant relations for centrifugal distortion constants.

Thus, there appear to be distinct advantages to determination of compliance constants, rather than force constants, for discussion of bonding and electronic structure. There is one apparent advantage to force constants over compliance constants. One can hope to calculate primary force constants from molecular orbital theory by calculation of energy as a function of the internal displacement coordinates, as $F_{kk} = (\partial^2 V/\partial R_k^2)_0$. The calculation of C_{kk} from molecular orbital theory is somewhat more involved. However, from satisfactory electronic orbitals it should certainly be possible, as outlined below:

Equation (1.13-8) can be rewritten as $2V = \sum_i f_i \sum_j C_{ij} f_j$. Then from (1.13-6) we have

$$2V = -\sum_i f_i S_i \tag{1.13-9}$$

Now let us apply a force f_k to coordinate k and allow all coordinates to adjust to a new equilibrium configuration (all $f_i = 0$ except f_k) with the constraint that

$$S_k = -C_{kk} f_k \tag{1.13-10}$$

From (1.13-9), $2V = -f_k S_k$ and combining this with (1.13-10) gives

$$C_{kk} = S_k^2/2V \tag{1.13-11}$$

Thus, to determine C_{kk}:

(1) Constrain coordinate k to displaced configuration:

$$\text{displacement} = S_k = -C_{kk} f_k$$

(2) Using molecular orbital theory, determine values of other S_i to give minimum energy with above constraint (1). This new minimum energy is $V_0 + V$.

(3) Calculate

$$\begin{aligned} C_{kk} &= S_k^2/2V \\ f_k &= -2V/S_k \\ C_{kl} &= -S_l/f_k = S_l S_k/2V \end{aligned}$$

Using this procedure, it should be possible to calculate both the diagonal and off-diagonal compliance constants provided we have a satisfactory molecular orbital description. Such a correlation of electronic structure theory with molecular vibrations would be extremely interesting.

In the following sections of this book, for many molecules compliance constants are given along with force constants. It is hoped that in future literature, vibrational spectroscopists will emphasize compliance constants at least as much as force constants.

REFERENCES

1. G. Herzberg, *Spectra of Diatomic Molecules*, Van Nostrand, Princeton, New Jersey, 1950, pp. 74, 75.
2. L. Pauling and E. B. Wilson, Jr., *Introduction to Quantum Mechanics*, McGraw-Hill, New York, 1935.
3. G. Herzberg, Spectra of Diatomic Molecules, Van Nostrand, Princeton, New Jersey, 1950, pp. 90–103.
4. D. H. Rank, D. P. Eastman, B. S. Rao, and T. A. Wiggins, *J. Opt. Soc. Am.* **52,** 1 (1962).
5. F. F. Cleveland, *Spectroscopia Mol.*, **14,** 32 (1965).
6. E. B. Wilson, Jr., J. C. Decius, and P. C. Cross, *Molecular Vibrations*, McGraw-Hill, New York, 1955.
7. G. Herzberg, *Infrared and Raman Spectra*, Van Nostrand, Princeton, New Jersey, 1945.
8. K. Nakamoto, *Infrared Spectra of Inorganic and Coordination Compounds*, John Wiley, New York, 1963.
9. J. C. Evans, *Infrared Spectroscopy and Molecular Structure* (Mansel Davies, ed.), Elsevier, Amsterdam, 1963, Chapter VI. See also R. S. Tobias, *J. Chem. Educ.*, **44,** 2–8, 70–79 (1967).

9a. L. H. Jones, *Inorg. Chem.*, **4,** 1472 (1965).

10. F. A. Cotton, *Chemical Applications of Group Theory*, Wiley-Interscience, New York, 1963.
11. H. H. Jaffé and M. Orchin, *Symmetry in Chemistry*, Wiley, New York, 1965.
12. W. Brügel, *An Introduction to Infrared Spectroscopy*, Wiley, New York, 1962.
13. J. H. Schachtschneider and R. G. Snyder, *Spectrochim. Acta*, **19,** 117 (1963).
14. Y. Morino and K. Kuchitsu, *J. Chem. Phys.*, **20,** 1809 (1952).
15. R. Gold, J. M. Dowling, and A. G. Meister, *J. Mol. Spectry*, **2,** 9 (1958).
16. B. L. Crawford, Jr. and J. Overend, *J. Mol. Spectry*, **12,** 307 (1964).
17. L. H. Jones, *Coord. Chem. Rev.* **1,** 351 (1966).
18. K. Nakamoto, *Infrared Spectra of Inorganic and Coordination Compounds*, Wiley, New York, 1963, pp. 231–235.
19. D. M. Dennison, *Rev. Mod. Phys.*, **12,** 175 (1940).
20. I. Karle and J. Karle, *J. Chem. Phys.*, **17,** 1052 (1949); S. J. Cyvin, *Spectrochim. Acta*, **10,** 828 (1959); S. J. Cyvin, *Molecular Vibrations and Mean Square Amplitudes*, Elsevier, Amsterdam, 1968.
21. E. B. Wilson, Jr., *J. Chem. Phys.*, **27,** 986 (1957).
22. D. Kivelson and E. B. Wilson, *J. Chem. Phys.*, **21,** 1229 (1953); **20,** 1575 (1952).
23. E. B. Wilson, Jr., *J. Chem. Phys.*, **5,** 617 (1937).

24. H. H. Nielsen, *Rev. Mod. Phys.*, **23,** 90 (1951).
25. J. H. Meal and S. R. Polo, *J. Chem. Phys.*, **24,** 1119, 1126 (1956).
25a. W. F. Edgell and R. E. Moynihan, *J. Chem. Phys.*, **27,** 155 (1957); **45,** 1205 (1966).
26. J. L. Duncan and I. M. Mills, *Spectrochim. Acta*, **20,** 523 (1964).
27. M. Iwasaki and K. Hedberg, *J. Chem. Phys.*, **36,** 594 (1962).
28. A. M. Mirri, F. Scappini, and P. G. Favero, *Spectrochim. Acta*, **21,** 965 (1965).
29. M. G. K. Pillai and S. G. Gnanadesikan, *Indian J. Pure Appl. Phys.*, **3,** 273 (1965).
30. H. C. Urey and C. A. Bradley, *Phys. Rev.*, **38,** 1969 (1931).
31. T. Schimanouchi, *J. Chem. Phys.*, **17,** 245 (1949).
32. J. L. Duncan, *J. Mol. Spectry.*, **18,** 62 (1965).
33. J. C. Decius, *J. Chem. Phys.*, **38,** 241 (1963).
34. R. S. Ottinger, Ph.D. Thesis, Oregon State University, Corvallis, Oregon, 1966.
35. R. R. Ryan, unpublished work, Los Alamos Scientific Laboratory, Los Alamos, New Mexico.
36. S. R. Polo, *J. Chem. Phys.*, **24,** 1133 (1956).
37. L. H. Jones and R. R. Ryan, *J. Chem. Phys.*, **52,** 2003 (1970).

Chapter 2

Potential Constants of Some Diatomic and Triatomic Molecules from Observed Data

2.1. Introductory Remarks about Units

For the calculation of force constants from observed vibrational frequencies, we are of course concerned with the units involved. For a diatomic molecule, for example, the force F is calculated from the relation

$$F = \mu\lambda \tag{2.1-1}$$

where μ is the reduced mass of the molecule, $m_1m_2/(m_1 + m_2)$, and $\lambda = 4\pi^2\nu^2$. If we insert μ in grams and ν in $\sec^{-1}$, the force constant F is given in dynes per centimeter, meaning the number of dynes of force tending to restore the molecule to its original length after it has been displaced from equilibrium by one centimeter (provided the bonding forces do not change during the process). However, it is convenient to express μ in atomic mass units and the frequency is generally observed as waves per centimeter (cm^{-1}). We, therefore, must divide μ in amu by Avogadro's number and multiply ν by the velocity of light in $cm \cdot sec^{-1}$.

The latest values (*1*) for these physical constants, for the new atomic weight scale with $M_{^{12}C} = 12.0000$ amu, are: Avogadro's number,

$$N_A = (6.02252 \pm 0.00009) \times 10^{23};$$

velocity of light,

$$C = (2.997925 \pm 0.000001) \times 10^{10} \text{ cm}\cdot\text{sec}^{-1}.$$

These values lead to the relation

$$F = (5.89146 \pm 0.00009) \times 10^{-2} \mu\nu^2 \quad \text{dyn cm}^{-1} \tag{2.1-2}$$

with μ in amu and ν in cm^{-1}. A more convenient unit for F is millidynes per Angstrom for which

$$F = 5.89146 \times 10^{-7} \mu\nu^2 \quad \text{mdyn Å}^{-1} \tag{2.1-3}$$

If one uses the old atomic weight scale, with $M_{^{16}O} = 16.0000$ amu, the conversion factor is slightly smaller (5.88957×10^{-7}).

The uncertainty given in Eq. (2.1-2) is only academic except in the few cases where ω_e is known very accurately. It leads to only a matter of $\pm 0.0015\%$ uncertainty in the force constant, whereas the frequency is generally not known to better than 0.01 to 1.0%. However, recently (*2*) ω_e for CO was found to be 2169.8232 cm^{-1}. With this frequency, application of Eq. (2.1-3), including the error limits of (2.1-2), leads to the force constant

$$F_{CO} = 19.02438 \pm 0.00029 \quad \text{mdyn Å}^{-1}$$

Thus, for the most accurate measurements to date, there is considerable uncertainty in the sixth figure due to uncertainty in physical constants.

For polyatomic molecules, the conversion factor is the same as in Eq. (2.1-3). Thus, when the force constant is in millidynes per Angstrom and the masses are expressed as atomic mass units (with $M_{^{12}C} = 12.000$), the eigenvalue λ is expressed as $5.89146 \times 10^{-7} \nu^2$ with ν in cm^{-1}.

In a quadratic valence force potential function of a polyatomic molecule, there are three specific types of force constants—pure stretching, stretch–bend, and pure bending. Thus, we will have terms like

$$2V = F_r(\Delta r)^2 + F_\alpha(\Delta\alpha)^2 + 2F_{r\alpha}(\Delta r)(\Delta\alpha) \tag{2.1-4}$$

The potential energy V is expressed in ergs $\times 10^{-11}$ if F_r is in millidynes per Angstrom and Δr is in Angstroms. With $\Delta\alpha$ in radians, F_α is then in mdyn Å rad^{-2} and $F_{r\alpha}$ is in mdyn·rad^{-1}. Some authors prefer to keep all force constants in the same unit (mdyn·Å^{-1} or dyn·cm^{-1}) by writing the potential function as

$$2V = (F_r\Delta r)^2 + \tilde{F}_\alpha(r_0\Delta\alpha)^2 + 2\tilde{F}_{r\alpha}(\Delta r)(r_0\Delta\alpha) \qquad (2.1\text{-}5)$$

where r_0 is $(r_i r_j)^{1/2}$ with r_i and r_j the equilibrium lengths of the bonds included by the angle α. If Eq. (2.1-5) is used, the symmetry coordinates involve the displacements $r_0\,\Delta\alpha$ rather than just $\Delta\alpha$ so the G matrix will be different than if (2.1-4) is used. The product of $F_\alpha G_\alpha$ and $\tilde{F}_\alpha\tilde{G}_\alpha$ will be the same. One must be cautious, however, in recognizing the difference in units for F_α and $\tilde{F}_\alpha$. Perhaps an even more appropriate unit for a bending constant is essentially an average of the two. Thus, if F_α in mdyn Å rad^{-2} is divided by $(r_1^0 r_2^0)^{1/2}$ Å rad^{-1}, the resulting force constant is $F_\alpha/(r_1^0 r_2^0)^{1/2} = H_\alpha$ mdyn rad^{-1}. This tells us the force required to deform the angle, α, by a small amount in radians from its equilibrium configuration.

We shall now consider some experimental results of vibrational spectra. In discussing these results it is important to consider what uncertainties exist in the reported force constants (or compliance constants or both) and to what extent the probable errors are interdependent. For this purpose, an estimate of the variance-covariance matrix is useful.

2.2. Force Constants of Diatomic Molecules

The vibrational frequencies ω_e have been determined for a large number of diatomic molecules. Herzberg (*3*) has compiled a list of this data. Most of it was obtained from electronic absorption spectra in which accompanying vibrational and rotational transitions were resolved. The force constants of these diatomic molecules are quite interesting both for comparison among themselves and for comparison with polyatomic molecules having similar bonds as part of their structure. Therefore, in Appendix V is presented a list of ω_e values [mostly from (*3*)] and the force constants calculated therefrom using Eq. (2.1-3). The anharmonicity constants $\omega_e x_e$ are given also as they are of interest for later discussion.

The order of listing in Appendix V is not alphabetical but according to position in the periodic table for convenient comparison. First, in Table V-1, are given homonuclear molecules grouped according to groups in the periodic table. Molecules made up of two different types of atoms within the same group are included in this table. Next come a large number of molecules, for which the two atoms are in different groups. These are arranged according to group in the periodic table for the more electropositive atom. Finally, in Table V-3, are given a number of hydrides, ordered by group of the nonhydrogen atom. For a diatomic molecule, the compliance constant is, of course, simply the reciprocal of the force constant.

2.3. Invariance of Force Constants to Isotopic Substitution

As mentioned in Sec. 1.11, the vibrational frequencies of isotopically substituted molecules generally give additional data for the calculation of force constants of polyatomic molecules. In order for this data to be useful, it is necessary to assume that force constants are invariant to isotopic substitution. Recently some very accurate measurements *(4)* on the vibration-rotation spectra of $H^{35}Cl$ and $D^{35}Cl$ showed that this is the case to a high degree of approximation. It was found *(4)* that $\omega_e = 2990.9742\ \mathrm{cm}^{-1}$ for $H^{35}Cl$ and $2145.1630\ \mathrm{cm}^{-1}$ for $D^{35}Cl$. From these values, with the use of Eq. (2.1-3), it is found that F or $k_e =$ 5.1629 mdyn Å^{-1} for $H^{35}Cl$ and 5.1630 for $D^{35}Cl$. This is within the accuracy of our knowledge of masses and Avogadro's number.*

This data on HCl, presented above, is by far the most accurate in the literature for evaluation of isotope effects on force constants. It gives us confidence that the force constant is invariant to isotopic substitution. For heavier molecules not involving hydrogen, there is even less likelihood of interaction of nuclear and electronic motion so that the force constants will indeed not depend on the isotopic masses. For lighter molecules, such as H_2, the data available at present is not accurate enought to determine the magnitude of such an isotopic effect on the force constant.

Some other isotopic pairs are given in Appendix V. For example, LiH and LiD, KH, and KD, and many other hydrides show reasonable agreement of force constants within pairs. However, there are a few pairs listed in Appendix V which have different calculated force constants, indicating that there is probably an error in the data. For example, BH–BD, NH–ND, and HgH–HgD. For the first two pairs, the ω_e are listed as uncertain and thus it is not surprising that discrepancies are noted in the force constants. However, the ω_e for HgH and HgD are given to several significant figures in *(3)*. The calculated force constants

* Actually when the Dunham corrections for vibration–rotation interactions are applied, both values are 5.1631 mdyn Å^{-1} [see J. L. Dunham, *Phys. Rev.*, **41**, 721 (1932)].

are 1.137 and 1.163 millidynes per Angstrom, respectively, so it is apparent that one or both of the ω_e assignments is in error. This would be suspected also from the fact that $\omega_e x_e$ for HgH (83.01) is not twice $\omega_e x_e$ for HgD (49.93) as it should be (*3*, p. 142).

2.4. Correlation of Force Constants with Positions of Atoms in the Periodic Table

It is interesting to compare the force constants of homonuclear diatomic molecules across the first period of the periodic table. Molecular orbital theory tells us what sort of bond order to expect for these molecules and the force constant should increase as the bond order increases. The electronic levels are discussed by Coulson (*5*) and by Herzberg [(*3*), pp. 344–346]. In Table 2.4-1 is given an approximate description of the

TABLE 2.4-1

Electronic Configuration and Force Constants of First Row Homonuclear Diatomic Molecules[a]

	σ_{1s}		σ^*_{1s}	σ_{2s}	σ^*_{2s}	σ_{2p}	π_{2p}	π^*_{2p}	σ^*_{2p}	B	A	Bond order	k_e	C_e
He_2^+	2		1							2	1	½	(3.12)	(0.321)
He_2	2		2							2	2	0	—	
Li_2^+		KK		1						1	0	½	—	
Li_2		KK		2						2	0	1	0.26	3.8
Be_2^+		KK		2	1					2	1	½	—	
Be_2		KK		2	2					2	2	0	—	
B_2^+		KK		2	2		1			3	2	½	—	
B_2		KK		2	2		11			4	2	1	3.58	0.279
C_2^+		KK		2	2	1[b]	11			5	2	1½	—	
C_2		KK		2	2	1	21			6	2	2	9.52	0.105
N_2^+		KK		2	2	1	22			7	2	2½	20.10	0.0498
N_2		KK		2	2	2	22			8	2	3	22.97	0.0435
O_2^+		KK		2	2	2	22	1		8	3	2½	16.59	0.0603
O_2		KK		2	2	2	22	11		8	4	2	11.77	0.0850
F_2^+		KK		2	2	2	22	21		8	5	1½	—	
F_2		KK		2	2	2	22	22		8	6	1	4.45	0.225
Ne_2^+		KK		2	2	2	22	22	1	8	7	½	—	
Ne_2		KK		2	2	2	22	22	2	8	8	0	—	

[a] The symbols used are B for bonding and A for antibonding electrons in valence shell. Units are mdyn Å^{-1} for force constants (k_e) and Å·mdyn^{-1} for compliance constants (C_e).

[b] It appears that σ_{2p} and π_{2p} are about the same energy for C_2^+ and C_2. [See (*3*), p. 34.]

electronic configuration for the first row, homonuclear diatomic molecules. The force constants are given, where known, in the last column. It is seen that there is a strong trend of increased force constant for increased bond order. However, we cannot relate force constant and bond order directly to each other. The bond order is only a nominal measure of bond strength. For example, N_2^+ and O_2^+ have the same bond order ($2\frac{1}{2}$) yet their force constants are appreciably different. O_2^+ has one more bonding and one more antibonding electron than N_2^+. The antibonding orbital detracts from the bond strength more than the bonding orbital adds.

TABLE 2.4-2

FORCE CONSTANTS (k_e) AND COMPLIANCE CONSTANTS (C_e) OF SECOND-ROW HOMONUCLEAR MOLECULES

Molecule	Bond order	k_e (mdyn Å^{-1})	C_e (Å·mdyn^{-1})
Na_2	1	0.17	5.9
Si_2	2	4.65	0.215
P_2	3	5.56	0.180
S_2	2	4.96	0.202
Cl_2^+	$1\frac{1}{2}$	4.29	0.233
Cl_2	1	3.29	0.304

[See (*5*), p. 98]. Any quantitative correlation of force constant with electronic structure must await further developments.

The second row follows the same trend, though much less dramatically, as seen in Table 2.4-2. The force constants of the alkali metals are surprisingly small, indicating that the σ_{2s} orbitals are not very good bonding orbitals.

In each group of the periodic table, the force constants of homonuclear diatomic molecules decrease with increasing atomic number. The larger atoms cannot get as close to each other and so do not form as strong a bond. The lead molecule, Pb_2, has a surprisingly large force constant for a bond between two heavy atoms.

Within a given group of the periodic table, the diatomic hydrides show decreasing force constant with increase in atomic number of the nonhydrogen atom. Furthermore, going across the table in a given period, the force constant of the hydride increases as the electronegativity of the nonhydrogen atom increases. This suggests that the hydride bonds have strong ionic character and thus their strength depends greatly on the electronegativity difference of the atoms.

Many other interesting correlations can be made going across and down the periodic table, among halides, oxides, etc. Of particular interest to coordination chemists are the molecules CO, CN, and NO as they frequently act as ligands to metals and the changes in their force constants can be correlated with the metal-ligand bonding. Their force constants will be discussed in the sections on metal-carbonyls and metal-cyanide complexes.

2.5. Force Constants of CO_2

The carbon dioxide molecule is linear and symmetric and thus has three fundamental frequencies in three different symmetry classes. This can be determined by using Eq. (1.4-1) with the internal coordinates. By convention, these are classified as ν_1, the symmetric CO stretching of Σ_g^+ symmetry; ν_2, the bending vibration of Π_u symmetry; and ν_3, the antisymmetric stretch of Σ_u^+ symmetry. The motions for the three normal vibrations are described in Appendix III. Since there is only one vibration in each symmetry class (counting a degenerate mode only once), the relative atomic motion is specifically described, as in Appendix III. Also since there is not more than one vibration in a given symmetry class, each vibration specifically determines a symmetry force constant. Thus, the symmetry force constants, $F_{CO} + F_{CO,CO}$, $F_{CO} - F_{CO,CO}$, and F_α, are readily determined from the three frequencies, ν_1, ν_3, and ν_2, respectively, using Eqs. (III-2)–(III-4) in Appendix III. Therefore, if the fundamental harmonic frequencies are known, one can readily calculate* F_{CO} (the CO stretching force constant), $F_{CO,CO}$ (the CO,CO interaction constant), and F_α (the OCO bending force constant).

For CO_2 a large number of vibrational levels have been observed. Courtoy (*6*) has made an extensive analysis of the observed levels and, using a relation such as Eq. (1.3-1), has calculated the various anharmonicity constants and the harmonic frequencies, ω_i. Since that time Amat and Pimbert (*7*) have pointed out that in Courtoy's analysis the observed isotopic shifts for the Fermi interaction constant do not agree with the calculated values. They suggest that the assignments of $2\nu_2$ and ν_1, as observed in the Raman spectrum, should perhaps be reversed. Gordon

* The linear symmetric *ABA* molecule is one of the few for which all the general quadratic force constants can be calculated from the fundamental frequencies of one species. Other examples are planar symmetrical X_3 and X_4 of D_3h and D_4h symmetry, respectively, and tetrahedral X_4.

and McCubbin (*8*) have followed the lead of Amat and Pimbert and have arrived at a new assignment which appears to be more consistent than Courtoy's. However, the problem is still open to question so the pertinent data of both references [(*6*) and (*8*)] are summarized in Table 2.5-1.

TABLE 2.5-1

VIBRATIONAL FREQUENCY DATA[a] FOR $^{12}C^{16}O_2$

	C (cm^{-1})	G & M (cm^{-1})
$\nu_1{}^b$	1388.17	1285.41
ν_2	667.40	667.38
$(2\nu_2)_0$	1285.40	1388.19
$(2\nu_2)_2$	1335.16	1335.13
ν_3	2349.16	2349.16
$(\nu_1)^{0c}$	1341.54	1332.87
$(2\nu_2)_0^0$	1332.04	1340.74
$\omega_1{}^d$	1354.91	1354.08
ω_2	673.00	672.96
ω_3	2396.49	2396.31

[a] Information in column C is from Courtoy (*6*) and in column G & M from Gordon and McCubbin (*8*). The ω_i under G & M were calculated from data in (*8*).

[b] These first five entries are observed frequencies. Original references are given in (*6*) and (*8*). The subscripts $_0$ and $_2$ on $2\nu_2$ refer to the value of *l*.

[c] The superscript ^{0} implies that these are the values of ν_1 and $2\nu_2$ which would pertain if Fermi resonance did not occur.

[d] These are the fundamental harmonic frequencies (for infinitesimal vibration).

Until the analysis is completely settled, we shall assume the true frequencies lie within these two sets of values. From these frequencies, we then calculate the harmonic force constants of Table 2.5-2 with the aid of Eqs. (III-10)–(III-12) in Appendix IIIB. The compliance constants are given also. They are elements of the force constant matrix inverse.

Thus, even for as simple a polyatomic molecule as CO_2 the harmonic force constants still have appreciable uncertainty. Of course, this arises primarily from the large Fermi resonance of ν_1 and $2\nu_2$ which complicates the analysis considerably.* Unfortunately in many other cases there is the possibility of Fermi resonance which leads to small frequency shifts but is difficult to recognize.

It is of interest to investigate the significance of the interaction constant, $F_{CO,C'O'}$. From Eq. (1.10-4) or (1.10-7) and the values of Table 2.5-2 it is calculated that $(CO_2)_{CO_1} = -1.263/16.025 = -0.0788 \pm 0.0004$. The same value is, of course, calculated from the compliance constants $[(CO_2)_{CO_1} = C_{CO,C'O'}/C_{CO} = -0.0788]$. This means that stretching one

TABLE 2.5-2

HARMONIC FORCE CONSTANTS AND COMPLIANCE CONSTANTS OF CO_2

F_{CO}	16.025 ± 0.006[a] mdyn Å^{-1}
$F_{CO,C'O'}$	1.263 ± 0.006[a] mdyn Å^{-1}
F_α	0.7854 ± 0.0002[b] mdyn Å rad^{-2}
C_{CO}	0.06279 ± 0.00002 Å mdyn^{-1}
$C_{CO,C'O'}$	-0.00495 ± 0.00002 Å mdyn^{-1}
C_α	1.2732 ± 0.0003 rad^2 Å^{-1} mdyn^{-1}

[a] The uncertainties for F_{CO} and $F_{CO,C'O'}$ arise from spanning the range of ω_i in Table 2.5-1.

[b] The uncertainty of F_α arises primarily from the uncertainty in bond length as given by Courtoy (*6*) ($R_{CO} = 1.1597_9$ Å at the equilibrium configuration).

CO bond a small amount tends to contract the other CO bond by 7.9% of the stretch of the first bond in order to minimize the energy.† This is readily interpretable on the basis of pi bond interaction. Thus, strong pi bonds are formed from the carbon to each oxygen. When one CO bond is stretched, its pi bond is weakened, freeing its pi electrons somewhat for stronger pi bond formation in the other bond, resulting in its contraction.‡ Of course, the CO′ sigma bond will also decrease when CO′ is stretching, leading to stronger attraction of the C for O″ and thus contraction of CO″. From consideration of change in overlap for sigma and pi

* For more detailed information on the analysis, the reader is referred to (*6*)–(*8*).

† It should be pointed out that this interaction displacement coordinate, $(CO_1)_{CO_2}$, does not describe the motion during a vibration but shows how CO_2 would *tend* to change to *minimize the energy* following a displacement of CO_1. It is an indication of the change in electron structure accompanying a displacement of CO_1.

‡ This has been discussed in some detail as the "resonance effect" by C. A. Coulson *et al.* in *Contributions to Molecular Spectra* (*38*).

bonds of CO, it appears (*9*) that the change in pi bonding is the dominant effect.

Cubic and quartic potential constants which are related to anharmonicity have also been calculated (*10–12*) for CO_2. However, these are not important for small displacements and are not considered further here. These higher order quantities are not necessary for calculation of harmonic force constants from the fundamental harmonic frequencies, though in many cases their inclusion has a significant effect on the calculated harmonic frequencies.

2.6. Force Constants of CS_2, CSe_2, CTe_2, and the Mixed *XCY* Species

Stoicheff (*13*) and Agar et al. (*14*) have observed and analyzed many bands of CS_2 and determined the anharmonic corrections and the fundamental harmonic frequencies. The observed fundamentals are given in Table 2.6-1 along with the calculated harmonic frequencies, force constants, and compliance constants (the potential constants were calculated by the author).

The Fermi resonance for $(2\nu_2)_0$ and ν_1 is much less important than in CO_2. However, the number of assigned levels is also much less than for CO_2 so there is appreciable uncertainty in the results. The "average" force constants are given at the end of Table 2.6-1 with the limits being arbitrarily determined by considering the two sets of results as extreme values.

It is interesting that these potential constants give an interaction displacement coordinate of $(CS_1)_{CS_2} = -0.083 \pm 0.002$ by application of Eq. (1.10-4). This is very close to the value for CO_2 suggesting that the bonding is quite similar.

For CSe_2 the data is less complete. Wentink (*15*) reports for the vapor

$$\nu_1 = 368 \pm 4 \quad \text{cm}^{-1}$$

$$\nu_2 = 308 \pm 4 \quad \text{cm}^{-1}$$

$$\nu_3 = 1303 \pm 2 \quad \text{cm}^{-1}$$

These are observed frequencies; no data is available for calculation of anharmonic corrections. However, a reasonable approximation can be made that the ratio of observed to harmonic fundamental frequencies is about the same for CSe_2 as for CO_2, CS_2, and OCS. This leads to the harmonic frequencies given in Table 2.6-2 for CSe_2.

TABLE 2.6-1

FUNDAMENTAL FREQUENCIES AND FORCE CONSTANTS FOR $C^{32}S_2$

		(cm⁻¹)		(cm⁻¹)
	ν_2	396.8	$(2\nu_2)_0$	802.1
	ν_1	658.0	ν_3	1532.5
	Stoicheff (*13*) (cm⁻¹)		Agar et al. (*14*) (cm⁻¹)	
ω_1	671.36		672.86	
ω_2	398.62		398.51	
ω_3	1551.92		1558.81	
F_{CS}	7.8292 mdyn Å^{-1}		7.8800 mdyn Å^{-1}	
$F_{CS,CS'}$	0.6608 mdyn Å^{-1}		0.6479 mdyn Å^{-1}	
F_α	0.5705 mdyn Å rad^{-2}		0.5701 mdyn Å rad^{-2}	
		Average		
	F_{CS}	7.86 ± 0.03 mdyn Å^{-1}		
	$F_{CS,CS'}$	0.65 ± 0.01 mdyn Å^{-1}		
	F_α	0.570 ± 0.001 mdyn Å rad^{-2}		
	C_{CS}	0.1282 ± 0.0003 Å mdyn^{-1}		
	$C_{CS,CS'}$	-0.0107 ± 0.0003 Å mdyn^{-1}		
	C_α	1.754 ± 0.003 rad^2 Å^{-1} mdyn^{-1}		

Using these values, we calculate for the force constants the values given in Table 2.6-2. The interaction displacement coordinate is then $(CSe_1)_{CSe_2} = -0.057 \pm 0.02$. This is somewhat lower than the analogous quantity for CO_2 and CS_2, though the uncertainty is too great to be sure if it is significantly lower.

Wentink (*15*) has observed the spectrum of some of the more unusual species such as OCSe, SCSe, and SCTe. For the linear XCY molecule with $X \neq Y$, as shown in Appendix III, Eq. (III-6), there are three stretching force constants in a General Quadratic Valence Force Field which determine two frequencies. Thus, the two stretching frequencies are not sufficient for calculation of the stretching force constants.

Wentink (*15*) has pointed out that the CO bond length is essentially constant in CO_2, OCS, and OCSe as is the CS bond length in CS_2, OCS,

TABLE 2.6-2

HARMONIC FREQUENCIES AND FORCE CONSTANTS OF CSe_2

(cm⁻¹)	
$\omega_1 = 373 \pm 6$	$F_{CSe} = 6.10 \pm 0.12$ mdyn Å^{-1}
$\omega_2 = 311 \pm 6$	$F_{CSe,CSe'} = 0.35 \pm 0.12$ mdyn Å^{-1}
$\omega_3 = 1323 \pm 9$	$F_\alpha = 0.47 \pm 0.03$ mdyn Å rad^{-2}

SCSe, and SCTe. This shows that the CX bond structure in XCY is not appreciably different for the different Y atoms. Jones (*16*) has shown that consideration of the changes in bonding accompanying bond stretches, as discussed for CO_2 in Sect. 2.5, leads to the conclusion that for these similarly bonded structures the interaction constant can be simply expressed as

$$F_{CX,CY} = \beta(F_{CX}F_{CY})^{1/2} \tag{2.6-1}$$

The proportionality constant, β, should be about the same for all ten species of the carbon dichalcogenides (CO_2, OCS, . . . , CTe_2). Because

TABLE 2.6-3

Fundamental Frequencies and Interaction Constants of the Linear Carbon Dichalcogenides[a]

Molecule	$F_{CX,CY}$ mdyn $Å^{-1}$	Calc. (cm^{-1}) ω_1	Calc. (cm^{-1}) ω_3	Obs. (cm^{-1}) ω_1	Obs. (cm^{-1}) ω_3
CO_2	1.263	1354.6	2396.6	1354.5	2396.4
CS_2	0.62	671	1560	672	1559
CSe_2	0.48	376	1308	373 ± 6	1323 ± 10
CTe_2	0.36	257	1121	—	—
OCS	0.88	889	2111	876 ± 6	2091 ± 10
OCSe	0.77	679	2070	650 ± 6	2045 ± 10
OCTe	0.67	567	2042	—	—
SCSe	0.54	515	1451	516 ± 6	1457 ± 10
SCTe	0.47	435	1385	431 ± 6	1386 ± 10
SeCTe	0.41	312	1222	—	—

[a] These values are calculated using $F_{CO} = 16.025$, $F_{CS} = 7.86$, $F_{CSe} = 6.1$, $F_{CTe} = 4.6$ mdyn $Å^{-1}$, and $\beta = 0.0788$.

of the similarity in bonding, it is also reasonable to assume that the primary stretching constants, F_{CO}, F_{CS}, etc., do not change appreciably among the different species. Thus, approximate values of the 20 stretching frequencies of the ten species can be calculated from five parameters (β, F_{CO}, F_{CS}, F_{CSe}, and F_{CTe}) with the aid of Eq. (2.6-1). The results, as taken from (*16*), are given in Table 2.6-3.

Equation (2.6-1) is based on two hypotheses (*16*) (or approximations) concerning the bonding. Hypothesis (1): For XCY, a change in bond order, α_{CX}, of the CX bond (due to stretching, for instance) leads to a proportional change in bond order of opposite sign, $-\beta\alpha_{CX}$, for the CY bond. The quantity β is constant for all X. Hypothesis (2): A change in bond length, Δr_{CX}, leads to a proportional change in bond order, α_{CX}, of the opposite sign and vice versa ($\Delta r_{CX} = -P_{CX}\alpha_{CX}$). The results of Table 2.6-3 indicate that these hypotheses give reasonable agreement with experiment. It is gratifying that the fourteen observed frequencies can be approximated fairly well using only five parameters (β, F_{CO}, F_{CS}, F_{CSe}, and F_{CTe}). The bending force constants for the species not involving Te are given in Table 2.6-4. These were calculated using Eq. (III-8) in Appendix III. It is quite interesting that the bending constants for the

TABLE 2.6-4

Bending Force Constants of Linear XCY

Molecule	ω_2 (cm^{-1})	F_α (mdyn Å rad^{-2})	(F_α/r_1r_2) (mdyn $Å^{-1}$)
CO_2	673.0	0.785	0.584
CS_2	398.6	0.570	0.236
CSe_2	311	0.47	0.16
OCS	530	0.67	0.37
OCSe	470	0.56	0.29
SCSe	358	0.53	0.20

mixed species are almost exactly the geometric mean of the bending constants for the two symmetric molecules involved; that is, $F_{XCY} \approx (F_{XCX} \cdot F_{YCY})^{1/2}$. From the last column of Table 2.6-4 using the values listed for CO_2, CS_2, and CSe_2, we calculate $F_{XCY} = 0.37$, 0.30, and 0.19 mdyn $Å^{-1}$ for OCS, OCSe, and SCSe, respectively. This is further indication of the similarity in bonding for this group of molecules.

2.7. Force Constants of XeF_2, KrF_2, and Related Molecules

Claassen et al. (*17*) have observed the spectrum of KrF_2 and compared it, along with calculated force constants, with XeF_2 (*18*). The results are given in Table 2.7-1. It is most interesting that F_{rr} has an opposite sign for KrF_2 and XeF_2. The interaction displacement coordinates [(from Eq. (1.10-4)] are $(XeF_1)_{XeF_2} = -0.046$ and $(KrF_1)_{KrF_2} = +0.081$. Thus, stretching one KrF bond leads to lengthening of the other KrF bond to minimize the energy. Though anharmonic corrections were not used, it is improbable that they would be large enough to explain this anomaly. Consideration of some similar molecules may be helpful in understanding this phenomenon.

A number of interhalogen compounds as well as HF_2^- and HCl_2^-, have electronic structures analogous to those of XeF_2 and KrF_2; that is, they have an excess of electrons over the number required to fill the valence shell p orbitals. A knowledge of their force constants, along with those of XeF_2 and KrF_2, should provide some insight into the nature of the unusual bonding forces which stabilize these molecules. The XY force constant and $XY,X'Y$ interaction constant are readily calculated from the symmetric

TABLE 2.7-1

OBSERVED FREQUENCIES AND CALCULATED FORCE CONSTANTS[a] AND COMPLIANCE CONSTANTS FOR XeF_2 AND KrF_2

	XeF_2	KrF_2
ν_1	515	449
ν_2	213.2	232.6
ν_3	558	588
F_r	2.84	2.46
$F_{rr'}$	0.13	−0.20
F_α/r_0^2	0.20	0.21
C_r	0.353	0.409
$C_{rr'}$	−0.016	0.033
$r_0^2C_\alpha$	5.0	4.8

[a] Units for ν are cm^{-1}, for F are mdyn $Å^{-1}$, and for C are Å $mdyn^{-1}$.

and antisymmetric stretching frequencies. Christe et al. (*19*) have suggested that this group of molecules can be separated into three subgroups according to their apparent bond order and the ratio of $F_{XY,X'Y}$ to F_{XY}, which is equivalent to the negative of the interaction displacement coordinate defined in Eq. (1.10-5). The literature results are summarized in Table 2.7-2. Actually it would be difficult to divide these results into three distinct groups; however, it is apparent that as the bond order increases, the bond–bond interaction becomes smaller.

The bonding in these molecules is believed (*20*, *21*) to involve primarily three p orbitals, one from each atom, and four electrons. Therefore, they have been referred to as three-center, four electron bonds (3c-4e). Two of these electrons are in a bonding orbital while two are in a nonbonding orbital. Therefore, the bond order should be less than one because there is only one bonding electron pair for two bonds. This agrees with the results of Table 2.7-2 though it is not possible to calculate n for XeF_2

or KrF_2. It seems to be agreed that for all of these XYX molecules, the X atoms are negatively charged. A consideration of resonance between valence bond structures is helpful.

$$\underset{\text{I}}{\bar{X}\cdots Y - X}, \quad \underset{\text{II}}{X - Y - X}, \quad \underset{\text{III}}{X - Y \cdots \bar{X}}$$

If we stretch the left hand XY bond, we favor structure I which approaches an X^- ion and an X–Y molecule*. Therefore if the XY molecule is

TABLE 2.7-2

FORCE CONSTANTS[a] FOR SOME LINEAR XYX MOLECULES

Molecule	F_{XY}	$F_{XY,X'Y}$	n[b]	$(X_2Y)_{X_1Y}$[c]	Ref.
HCl_2^-	0.80	0.61	0.16	−0.76	(*19*)
HF_2^-	2.31	1.72	0.24	−0.74	(*19*)
Cl_3^-	0.96	0.55	0.30	−0.57	(*19a*)
$BrCl_2^-$	1.02	0.47	0.36	−0.46	(*19a*)
Br_3^-	0.91	0.32	0.38	−0.36	(*19b*)
ICl_2^-	1.00	0.36	0.42	−0.36	(*19*)
IBr_2^-	0.91	0.30	0.42	−0.33	(*19b*)
ClF_2^-	2.35	0.17	0.55	−0.07	(*19*)
XeF_2	2.84	0.13	—	−0.05	(*18*)
KrF_2	2.46	−0.20	—	+0.08	(*17*)

[a] Units of force constants are millidynes per Angstrom.
[b] n is the ratio of F_{XY} for XYX to F_{XY} for the diatomic molecule XY. It is a measure of bond order.
[c] This is the interaction displacement coordinate, defined by Eq. (1.10-4).

much shorter than the XY distance in XYX, the interaction coordinate, $-(X_1Y)_{X_2Y}$, will be large. So as n decreases, $-(X_1Y)_{X_2Y}$ should increase as is observed. Perhaps the KrF molecule would be less stable than the KrF bond in KrF_2; this would explain the positive value for $(KrF_2)_{KrF_1}$. This is merely an attempt at a simple explanation of the data and does not imply that it is a conclusive interpretation.

Coulson (*22*) has offered an explanation of the negative value for F_{rr} of KrF_2 based on consideration of resonance interaction with a "no-bond" structure. He considers that there are three important resonance structures to consider

$$\underset{\text{(I)}}{F^- - X^+ \quad F} \qquad \underset{\text{(II)}}{F \quad X^+ - F^-} \qquad \underset{\text{(III)}}{F \quad X \quad F}$$

with X representing the rare-gas atom. For a molecule of reasonable

* Actually, for XeF_2 (and KrF_2) we should say an X^- ion and an $X - Y^+$ ion.

stability, structure III is relatively unimportant and we only have to consider structures I and II. If we stretch the left-hand bond from its equilibrium configuration, this will tend to reduce the weight of structure I and increase that of II, leading to a small decrease in length of the other X–F bond and thus a positive interaction constant [$(XF_1)_{XF_2}$ being negative]. However, for a molecule of small stability, such as KrF_2, the "no-bond" structure III becomes important. Then if the left hand XF bond is stretched, the contribution from structure III (which implies extra long XF bonds) is increased and the effect of structure II is lessened, or perhaps overwhelmed, resulting in a net small decrease or perhaps even an increase in the right-hand X–F bond. By such arguments Coulson (*22*) suggests that the relatively unstable molecule, KrF_2, should have a much smaller interaction constant, F_{rr}, than XeF_2, and it may even be negative, as found experimentally.

This is an example of the sort of interesting bonding information one can obtain from studies of vibrational spectra. However, the picture presented above is no doubt greatly oversimplified and certainly not unique. Further discussion of the bonding can be found in (*20*), (*21*), and (*23*).

2.8. The Azide Ion

The azide ion is believed to be linear and symmetric. The equations relating force constant to frequency are particularly simple as obtained from (III-10)–(III-12) in appendix III by equating M_A and M_B

$$\begin{aligned}\lambda_1 &= (F_{\mathrm{NN}} + F_{\mathrm{NN}',\mathrm{NN}''})(1/M_{\mathrm{N}})\\ \lambda_3 &= (F_{\mathrm{NN}} - F_{\mathrm{NN}',\mathrm{NN}''})(3/M_{\mathrm{N}})\\ \lambda_2 &= (F_\alpha/r^2)(6/M_{\mathrm{N}})\end{aligned} \tag{2.8-1}$$

The vibrational spectra of the alkali azides have been studied in the solid state by Papazian (*24*) and Bryant (*25*, *26*). The frequencies vary considerably with cation. Therefore, for purposes of force constant calculations, the aqueous solution frequencies are desirable. These have been observed by the author (*27*) and found to occur as listed in Table 2.8-1. Anharmonic corrections are not known, though estimates of some of them can be made from the solid state spectrum (*26*). There is also a possibility of some Fermi resonance of $2\nu_2$ with ν_1. Allowing for these uncertainties, estimates of harmonic frequencies have been made by the author as they appear in Table 2.8-1. The calculated force constants and compliance constants are given in Table 2.8-1 for both the observed and the harmonic frequencies. It is seen that though the bond stretching force constant is considerably weaker than for isoelectronic CO_2, the stretch–stretch interaction is definitely greater.

2.9. POTENTIAL CONSTANTS OF N_2O, NCO, AND NCS

Several isotopic species of N_2O have been investigated with vibrational spectroscopy. A very thorough high resolution analysis of $^{14}N^{14}N^{16}O$

TABLE 2.8-1

FUNDAMENTAL FREQUENCIES, FORCE CONSTANTS, AND COMPLIANCE CONSTANTS FOR N_3^-

	Observed	Harmonic (Estimated)
ν_1	1343	1347 ± 10 cm^{-1}
ν_2	637.5	642 ± 10
ν_3	2048	2065 ± 10
F_{NN}	13.21	13.35 ± 0.12 mdyn Å^{-1}
$F_{NN',NN''}$	1.67	1.62 ± 0.12 mdyn Å^{-1}
F_α	0.74	0.75 ± 0.02 mdyn Å rad^{-2}
C_{NN}		0.0760 ± 0.0006 Å mdyn^{-1}
$C_{NN',NN''}$		−0.0092 ± 0.0006 Å mdyn^{-1}
C_α		1.35 ± 0.04 rad^2 Å^{-1} mdyn^{-1}
Interaction coordinate, $(NN')_{NN''} = -0.12$		

has been reported by Pliva (*28*). He has determined the harmonic frequencies ω_1, ω_2, and ω_3, the centrifugal distortion constant, D_e, the Coriolis zeta values, and other molecular constants. The three frequencies, along with D_e, are sufficient data to calculate the four quadratic compliance constants, C_{NN}, C_{NO}, $C_{NN,NO}$, and C_α. However, inclusion of ω_2 and D observed for other isotopic species (*29*, *30*) permits a least squares calculation of the C_{ij} using the Ottinger (*42*)–Ryan (*42a*) computer program. This is discussed herein in some detail to give an indication of the sort of information we can obtain from such a treatment.

In treating the vibrational frequency and centrifugal distortion data, one is confronted with the problem of how to weight the various data in the least squares procedure. The experimental uncertainty in the data should be included in the weights. The program minimizes $\Sigma(\Delta\varphi_i)^2$ with $\varphi_i = \frac{1}{4}\pi^2\nu_i^2$, and $\Delta\varphi_i = \varphi_i\,(\text{obs}) - \varphi_i\,(\text{calc})$. The weights should be

TABLE 2.9-1

OBSERVED AND CALCULATED VIBRATIONAL DATA FOR ISOTOPIC SPECIES OF NITROUS OXIDE: $^iN^jN^kO$

i-j-k →	14-14-16	15-14-16	15-15-16	15-15-18	14-15-16	14-14-18	15-14-18	14-15-18
ω_1 obs.	1298.65 ± 0.1[a]	1282.0 ± 1.0[b]	1280.0 ± 2.5[b]	1239.7 ± 2.5[c]				
ω_1 calc.	1298.66	1282.0	1280.5	1236.9	1296.7	1255.3	1238.9	1252.8
ω_3 obs.	2282.0 ± 0.1[a]	2258.6 ± 0.4[b]	2209.5 ± 0.4[b]	2201.5 ± 0.4				
ω_3 calc.	2282.0	2258.7	2209.0	2202.1	2233.1	2275.6	2251.8	2226.9
$D_e \times 10^7$ obs.	1.719 ± 0.01[a]	1.65 ± 0.1[b]	1.66 ± 0.1[b]	1.43 ± 0.15[c]				
$D_e \times 10^7$ calc.	1.722	1.598	1.601	1.434				
Σ_{N-N}, Å^2	0.001121	0.001103	0.001104	0.001083	0.001083	0.001120	0.00110	0.001103
Σ_{NO}, Å^2	0.001333	0.001332	0.001313	0.001312	0.001272	0.001293	0.001292	0.001273
$\Sigma_{N\ldots O}$, Å^2	0.001698	0.001672	0.001670	0.001622	0.001697	0.001651	0.001623	0.001650
ζ_{21} obs.	±0.212							
ζ_{21} calc.	±0.2006					±0.2041		
ζ_{23} obs.	±0.977							
ζ_{23} calc.	±0.9797					±0.9790		
ω_2 obs.	596.36 ± 0.04[a]	592.79[d]	579.02[d]					
ω_2 calc.	596.36	592.81	579.09	574.24	582.73	591.65	588.07	577.91

$C_{NN} = 0.05443 \pm 0.00025^e$ Å mdyn^{-1} | $F_{NN} = 18.48 \pm 0.09^e$ mdyn Å^{-1}

$C_{NO} = 0.08506 \pm 0.00042$ Å mdyn^{-1} | $F_{NO} = 11.83 \pm 0.06$ mdyn Å^{-1}

$C_{NN,NO} = -0.00522 \pm 0.00019$ Å mdyn^{-1} | $F_{NN,NO} = 1.13 \pm 0.04$ mdyn Å^{-1}

$C_\alpha = 1.5019 \pm 0.0002$ rad^2 Å^{-1} mdyn^{-1} | $F_\alpha = 0.6658 \pm 0.0001$ mdyn Å rad^{-2}

$(NN)_{NO} = -0.0614 \pm 0.0017$ | $(NO)_{NN} = -0.0959 \pm 0.0028$

[a] From (*28*). Uncertainties are estimated from the range of results given for different initial assumptions.

[b] From (*29*) using anharmonic corrections of Pliva (*28*) corrected for isotope shifts. Uncertainties are estimates made by the author based on observed ν_1^0 and ν_3 and the anharmonic corrections of Pliva (*28*) corrected for isotope shifts.

[c] From (*30*) and anharmonic corrections of Pliva (*28*) corrected for isotope shifts. The rather large uncertainty is given to ω_1 because the Product Rule is not obeyed for ω_1 and ω_3 of the 14-14-16 species.

[d] These are calculated from the observed values of ν_2 (*28a*) and the anharmonic corrections of Pliva (*28*) corrected for isotope shifts.

[e] Uncertainties given are least squares standard deviations.

proportional to $1/(d\varphi_i)^2 = k\nu_i^6/(d\nu_i)^2$. Therefore the program weights each eigenvalue, φ_i, by ν_i^6 and it is appropriate to further weight each φ_i by $1/(d\nu_i)^2$ with $d\nu_i$ equal to the experimental uncertainty in ν_i.

For the distortion constants we wish to minimize $\Sigma P_i(\Delta D_i)^2$, where $\Delta D_i = D_i\,(\text{obs}) - D_i\,(\text{calc})$. The weighting factor P_i is taken as $1/(dD_i)^2$ with dD_i equal to the estimated experimental uncertainty in the centrifugal distortion constant D_i. A further constant factor is required to weight the centrifugal distortion constants relative to the frequencies. Without specific information to the contrary, it is appropriate to choose this constant factor so that the sums of the weighted squares of the residuals of the frequency eigenvalues divided by the number of observed frequencies equals the sum of the weighted squares of the residuals of the distortion constants divided by the number of observed distortion constants: that is, $\Sigma P_i(\Delta D_i)^2/n_D = \Sigma P_j(\Delta\varphi_j)^2/n_\phi$. This gives equal weight, on the average, to frequency data and centrifugal distortion data. This is an approximation, of course, but large relative changes in the weights must be made in order to cause significant changes in the compliance constant estimates.

After determining the best estimates for the compliance constants, it is easy to calculate the interaction coordinates from Eq. (1.10-7), $(l)_k = C_{lk}/C_{kk}$, where $(l)_k$ represents the change in internal coordinate l required to minimize the energy if k is constrained to a unit positive displacement.

Table 2.9-1 lists the results of the above procedure. It is interesting to investigate the sensitivity of various types of data to the force constant solution. For this purpose the frequencies, centrifugal distortion constants, amplitudes, and Coriolis constants were calculated from a set of compliance constants having $C_{NN,NO}$ fixed at -0.001 Å mdyn and C_{NN} and C_{NO} calculated to fit ω_1 and ω_3 of $^{14}N^{14}N^{16}O$. The results are presented in Table 2.9-2.

The calculations recorded in Table 2.9-2 show that some of the isotopic frequency shifts and the Coriolis constant ζ_{21} for this molecule are quite a sensitive measure of potential constants. Thus, the $^{14}N^{15}N^{18}O$ molecule shows very large dependence of frequency on force constant solution, many times the accuracy of measurement in this case. On the other hand, the isotope shift for the molecule $^{15}N^{14}N^{18}O$ would be relatively useless for estimation of compliance constants. For all isotopic species, the Coriolis constant ζ_{21} is also quite sensitive to the force constant solution.

It is apparent that the centrifugal distortion constant is not very sensitive to the potential constant solution for a molecule such as N_2O;

however, the fact that this constant can be measured quite accurately means that it is useful for estimation of the potential constants. The mean square amplitude of the N—O bond (which is the amplitude most sensitive to force constants) would have to be known within 5% or less in order to be useful for calculating the compliance constants. For N_2O this would be out of the question because the NN and NO distances are within a few hundredths of an Angstrom of each other [1.127 and 1.186 Å, respectively (*30*)]. Even for NO_2, which has been studied quite thoroughly by electron diffraction, the mean square amplitudes have standard deviations greater than 10% (*42*). The calculated amplitudes may, of course, be useful for evaluating electron diffraction results.

TABLE 2.9-2

SENSITIVITY[a] OF VIBRATIONAL DATA TO FORCE CONSTANTS FOR NITROUS OXIDE

	14-14-16	15-14-16	15-15-16	15-15-18	14-15-16	14-14-18	15-14-18	14-15-18
$\Delta\omega_1$ (cm^{-1})	0	4.38	3.00	−1.93	−1.68	−4.50	−0.31	−6.40
$\Delta\omega_3$ (cm^{-1})		−7.75	−5.20	3.43	2.90	8.13	0.57	11.30
$\Delta D_e \times 10^7$ (cm^{-1})	0.046	0.049	0.048	0.034				
$\Delta\Sigma_{N-N}$ ($Å^2$)	0.0008[b]							
$\Delta\Sigma_{N-O}$ ($Å^2$)	−0.00011[b]							
$\Delta\Sigma_{N\ldots O}$ ($Å^2$)	0.0004[b]							
$\Delta\zeta_{21}$	±0.153					±0.144		
$\Delta\zeta_{23}$	±0.019					±0.019		

[a] The Δ values given are the calculated shifts from the compliance constant solution of Table 2.9-1 to the compliance constants, $C_{NN,NO} = -0.001$, $C_{NN} = 0.06436$, and $C_{NO} = 0.07152$ Å $mdyn^{-1}$.
[b] These values are the same for all isotopic species above.

The correlation coefficients are given in Table 2.9-3 for the compliance constants and for the interaction coordinates. This table shows that there is an extremely high correlation among these results; thus, a large positive error in C_{NN} is probable only if there is a large negative error in C_{NO} or a large positive error in $C_{NN,NO}$ or both.

The interaction coordinates, given as $(NN)_{NO}$ and $(NO)_{NN}$ in Table 2.9-1, give us interesting information about the electronic structure. The N_2O molecule can be considered as a distorted CO_2 molecule. In a discussion of CO bonds of metal carbonyls, Kettle (*46*) reports that small changes in CO distances should cause relatively little change in sigma overlap but significant change in pi overlap. One would expect a similar

TABLE 2.9-3

CORRELATION COEFFICIENTS[a] FOR COMPLIANCE CONSTANTS AND INTERACTION COORDINATE OF N_2O

	C_{NN}	C_{NO}	$C_{NN,NO}$
C_{NN}	1.00	−1.00	1.00
C_{NO}		1.00	−1.00
$(NN)_{NO}$ with $(NO)_{NN}$, $\rho = -0.97$			

[a] The correlation coefficient, ρ_{ij}, of x_i with x_j is equal to COV $(x_i, x_j)/\sigma(x_i)\sigma(x_j)$ where COV stands for covariance and $\sigma(x_i)$ are the standard deviations of x_i.

argument to hold for the bonds in N_2O for which the NN pi bonding should be somewhat greater and the NO pi bonding somewhat less than that for CO in CO_2. A stretch of NN will release a portion of the central N pi orbitals for NO pi bonding, thus leading to a decrease in NO bond length for minimum energy. A stretch of NO will similarly tend to decrease NN; however the effect will be smaller as the NO bond is weaker to start with than is the NN bond. It is quite encouraging that the average of $(NN)_{NO}$ and $(NO)_{NN}$ is equal to −0.079 which is identical to the value found (*16*) for $(CO_2)_{CO_1}$ in CO_2, in accord with the expected similarity in bonding.

Table 2.9-4 lists the force constants for the cyanate ion, NCO^-, in a KBr lattice as given by Maki and Decius (*31*), and the force constants for two other isoelectronic species, CO_2 and N_3^-, as discussed previously, for comparison with N_2O.

One can correlate the primary force constants with the three canonical resonance structures

$$:A \equiv B - \ddot{\underset{\cdot\cdot}{C}}: \qquad :\dot{\underset{\cdot}{A}} = B = \ddot{\underset{\cdot\cdot}{C}} \qquad :\ddot{\underset{\cdot\cdot}{A}} - B \equiv C:$$

I II III

TABLE 2.9-4

FORCE CONSTANTS OF SOME LINEAR TRIATOMIC MOLECULES

A-B-C	F_{AB}	F_{BC}	$F_{AB,BC}$	F_α
OCO	16.03	16.03	1.26	0.79
NNN^-	13.4	13.4	1.6	0.75
NNO	18.5	11.8	1.1	0.67
NCO^-	15.9	11.0	1.4	0.73

For CO_2 we expect considerable contribution from each of the three structures I, II, and III. For N_3^- structures, I and III involve a double negative charge on one of the end atoms which decreases their stability. This reduces the resonance stabilization and thus leads to weaker bonds. For N_2O structure III places a double negative charge on the end N atom and positive charges on the O and central N atoms; thus the primary resonance structures to consider are I and II, leading to perhaps 50% triple bond character in the N—N bond and 50% single bond character in the NO bond. Thus we have a high N—N force constant and a

TABLE 2.9-5

Interaction Displacement Coordinates for Some Linear Triatomic Molecules

Molecule		
OCO	$(CO_1)_{CO_2} = -0.079$	
NNN⁻	$(N_1N_2)_{N_2N_3} = -0.12$	
NNO	$(NN)_{NO} = -0.061$ $(NO)_{NN} = -0.096$	−0.079 av
NCO⁻	$(NC)_{CO} = -0.088$ $(CO)_{NC} = -0.13$	−0.11 av
NCS⁻	$(NC)_{CS} = -0.056$ $(CS)_{NC} = -0.17$	−0.11 av

low NO force constant. The same argument applies for NCO^-, though structure III is not quite as unstable as the carbon atom does not have the formal positive charge. The discussion of change in pi overlap given for N_2O earlier in this section will lead to similar conclusions.

For the thiocyanate ion, harmonic force constants have been determined (*32*) with the aid of isotopic substitution of ^{15}N and ^{13}C. The results are $F_{CN} = 16.0$, $F_{CS} = 5.2$, $F_{CN,CS} = 0.9$ mdyn Å^{-1}, and $F_\alpha = 0.58$ mdyn Å rad^{-2}, with rather large limits of error. It is seen that the CN bond strength is about the same in NCO^- and NCS^-.

The interaction displacement coordinates Eq. (1.10-4) for these molecules are given in Table 2.9-5. The negatively charged molecules appear to have somewhat greater interaction displacement coordinates than the neutral molecules. It is quite interesting that the negative ions have the same average interaction displacement coordinates, somewhat greater than the value for CO_2 and N_2O. This encourages the idea that we can correlate interaction coordinates with bonding.

2.10. Force Constants of HCN

The linear HCN molecule has been the subject of numerous infrared spectral investigations. Rank et al. (*33*) have listed a large number of vibrational levels and calculated anharmonicity constants and fundamental harmonic frequencies. Allen et al. (*34*) made new observations on HCN and DCN and calculated vibrational constants from the results. There are some differences in the two results. Suzuki et al. (*35*) have treated the

TABLE 2.10-1

Vibrational Data[a] of HCN and DCN

	HCN	DCN
	(cm^{-1})	(cm^{-1})
ν_1	2096.7	1925.3
ν_2	712.1	569.1
ν_3	3311.5	2630.3
ω_1	2132.0	1935.5
ω_2	726.5	579.4
ω_3	3438.5	2703.6
	$F_{CH} = 6.23$ mdyn Å^{-1}	$F_{CH,CN} = -0.216$ mdyn Å^{-1}
	$F_{CN} = 18.77$ mdyn Å^{-1}	$F_\alpha = 0.259$ mdyn Å rad^{-2}

[a] From (*35*).

raw data given in (*33*) and (*34*) and have arrived at a different set of vibrational constants. The differences arise because of a difference in the importance attached to various anharmonicity constants. The results of Suzuki et al. (*35*) are the most consistent and are to be preferred at present. The observed fundamentals and the calculated harmonic frequencies are given in Table 2.10-1 along with force constants calculated from the harmonic frequencies using Eqs. (III-6) and (III-7) for the two isotopic species.

Uncertainties in the results of Table 2.10-1 are not known. The biggest discrepancy is the assignment by Rank et al. (*33*) of 2119.9 cm^{-1} to the harmonic frequency ω_1 of HCN. This leads to a value of about 18.56 for F_{CN} rather than 18.77. For the present, the values listed in Table 2.10-1 are to be preferred as they are consistent with the use of one set of force constants for HCN and DCN.

The negative value for the interaction constant $F_{CH,CN}$ is somewhat surprising. From Eq. (1.10-4) and the constants of Table 2.10-1, it is found that the interaction displacement coordinates (see Sect. 1.10) are $(CN)_{CH} = 0.012$ and $(CH)_{CN} = 0.035$. Thus, stretching the CH bond or the CN bond results in a slight extension of the other bond for minimum energy. This can be explained by consideration of the resonance structures below:

$$\underset{\text{I}}{\mathrm{H{-}C{\equiv}N\!:}} \qquad \underset{\text{II}}{\overset{+}{\underset{\leftarrow}{\mathrm{H}}}\ \mathrm{:\bar{C}{\equiv}N\!:}} \qquad \underset{\text{III}}{\overset{+}{\underset{\leftarrow}{\mathrm{H}}}\ \mathrm{:C{=}\ddot{\bar{N}}\!:}}$$

Stretching of the CH bond increases the contributions of structures II and III, which combination leads to a weaker and thus longer CN bond. Or, conversely, if CN is stretched, structure III is favored leading to a longer CH bond for minimum energy.

2.11. The Water Molecule

The water molecule serves as a much-studied example of a bent symmetrical triatomic molecule. The point group symmetry is C_{2v} and as a result of the angle being less than 180° the vibrational treatment is more complicated than for the linear triatomic molecule. For the linear molecule in the quadratic approximation, there was no interaction between bond-stretching and angle-bending; this is not true for the bent triatomic molecule. The frequencies of nonlinear symmetric YXY are related to the force constants by the expressions

$$\lambda_1 + \lambda_2 = \left(\frac{1}{m_y} + \frac{1 + \cos\alpha}{m_x}\right)(F_r + F_{rr}) + \frac{2}{r^2}\left(\frac{1}{m_y} + \frac{1 - \cos\alpha}{m_x}\right)F_\alpha - \frac{\sqrt{2}\sin\alpha}{rm_x}F_{r\alpha} \tag{2.11-1}$$

$$\lambda_1\lambda_2 = \frac{2}{r^2}\left(\frac{1}{m_y^2} + \frac{2}{m_x m_y}\right)[(F_r + F_{rr})F_\alpha - F_{r\alpha}^2] \tag{2.11-2}$$

$$\lambda_3 = \left(\frac{1}{m_y} + \frac{1 - \cos\alpha}{m_x}\right)(F_r - F_{rr}) \tag{2.11-3}$$

where

λ_i is $4\pi^2\nu_i^2c^2$, with ν_i the frequency of vibration i in cm^{-1},
r is the equilibrium XY bond length,
m_k are the masses, and
α is the YXY angle at equilibrium.

The form of the vibrations is given in Fig. 2.11-1. Actually since there are two A_1 vibrations, the actual motion of the normal modes cannot be given without a normal coordinate treatment. However, the normal modes for ν_1 and ν_2 are close to the schematic motions shown under ν_1 and ν_2 in Fig. 2.11-1. The vibration ν_1 will involve a small amount of the motion shown for ν_2 and vice versa.

Benedict et al. (36) have treated the spectra of gaseous H_2O and D_2O and have arrived at values for the fundamental harmonic frequencies.

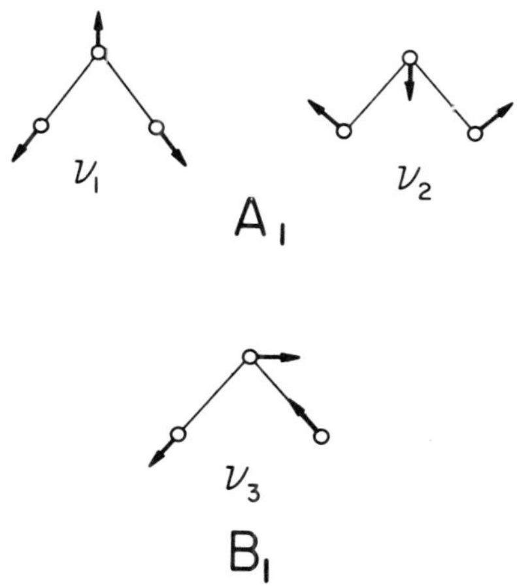

Fig. 2.11-1 Schematic representation of normal modes of H_2O.

These are slightly different than earlier values and are perhaps the best to use for force constant calculations. However, their value for ω_1, given as 2783.8, should apparently be 2763.8, as pointed out by Kuchitsu and Bartell (*37*). The latter authors have calculated the force constants for H_2O and give a set of harmonic frequencies close to the values of Benedict et al. (*36*) with the exception of ω_1. The observed fundamentals, harmonic frequencies, and force constants for H_2O and D_2O are given in Table 2.11-1. The force constants given here were calculated from the symmetry force constants given in (*37*).

The negative stretch–stretch interaction constant for the water molecule has been the subject of considerable discussion, particularly by Coulson et al. (*38*) and Heath and Linnett (*39*). The arguments of (*38*) and (*39*) for the sign of F_{rr} are in contradiction to each other. Before considering this further, however, it is appropriate to consider other force fields. The force constants of Table 2.11-1 are for a General Quadratic Valence

TABLE 2.11-1

VIBRATIONAL FREQUENCIES[a] AND FORCE CONSTANTS OF WATER VAPOR

	H_2O	D_2O	
ν_1[b]	3656.65	2671.46	
ν_2	1594.59	1178.33	$F_r{}^d = 8.45_4$ mdyn Å^{-1}
ν_3	3755.79	2788.05	$F_{rr} = -0.10$ mdyn Å^{-1}
			$F_{r\alpha} = 0.21_8$ mdyn rad^{-1}
ω_1[c]	3832.0	2764.0	$F_\alpha = 0.69_7$ mdyn Å rad^{-2}
ω_2	1648.9	1206.1	
ω_3	3942.6	2888.8	

[a] Values are in cm^{-1}.
[b] The ν_i are observed values from (*36*).
[c] The ω_i are calculated values from (*37*).
[d] Though the uncertainties in the ω_i are not known, the author's evaluation of the literature indicates that the third decimal place in the force constants is uncertain.

Force Field. Two other commonly discussed force fields are the General Quadratic Central Force Field (CQCFF) and the Urey–Bradley Force Field (UBFF). The three potential functions are given below:

GQVFF:

$$2V = F_r[(\Delta r_1)^2 + (\Delta r_2)^2] + F_\alpha(\Delta\alpha)^2 + 2F_{rr}(\Delta r_1)(\Delta r_2) + 2F_{r\alpha}(\Delta\alpha)(\Delta r_1 + \Delta r_2)$$

GQCFF:

$$[2V = K_r(\Delta r_1)^2 + (\Delta r_2)^2] + K_Q(\Delta q)^2 + 2K_{rr}(\Delta r_1)(\Delta r_2) + 2K_{rq}(\Delta q)(\Delta r_1 + \Delta r_2)$$

UBFF:

$$2V = K'r_0(\Delta r_1 + \Delta r_2) + H'r_0^2(\Delta\alpha) + F'q_0(\Delta q) + K(\Delta r_1^2 + \Delta r_2^2) + Hr_0^2(\Delta\alpha)^2 + F(\Delta q)^2$$

In the above expressions r_1 and r_2 are the two OH distances, α is the HOH angle, q is the H–H distance, and r_0 and q_0 are the equilibrium OH and HH distances, respectively. The constants K' and H' are not independent, they are related to F'.

Each of the three potential functions is completely general in the quadratic approximation and thus they must be equivalent with merely a different emphasis on the displacement coordinates. The calculated force constants are given in Table 2.11-2.

Several interesting observations can be made. For the three potential functions, the OH stretching constants are different. This, of course, is because they are defined differently. For valence forces, F_r tells us the force required to stretch one OH bond while the other OH bond and the angle remain fixed. For central forces, K_r tells us the force required to stretch one OH bond while the other OH bond and the H–H distance remain fixed. For Urey–Bradley forces, K does not carry such a simple definition. It is not an easily definable force constant in itself, because of the redundancy, but is related to F_r through the expression $K = F_r - F \sin^2(\alpha/2) - F' \cos^2(\alpha/2)$ [see (*16*) of Chapter 1]. Any reasonable potential field between the two hydrogen atoms requires that F' be negative and much smaller in absolute value than F for the Urey–Bradley Field. Thus the UBFF is not a reasonable potential function for H_2O.

TABLE 2.11-2

FORCE CONSTANTS[a] OF H_2O FOR THREE DIFFERENT GENERAL QUADRATIC POTENTIAL FUNCTIONS

GQVFF	GQCFF	UBFF
$F_r = 8.45$	$K_r = 9.13$	$K = 8.26$
$F_\alpha = 0.70$	$K_p = 2.03$	$Hr_0^2 = 0.90$
$F_{rr} = -0.10$	$K_{rr} = 0.58$	$F = 0.075$
$F_{r\alpha} = 0.22$	$K_{rq} = -1.23$	$F' = 0.40$

[a] The units are millidynes per Angstrom for all stretching constants. For the bending constants, F_α and Hr_0^2, the units are millidynes·Angstroms·radians^{-2}. For $F_{r\alpha}$ the units are millidynes·radians^{-1}.

More striking is the difference in the interaction constants, F_{rr} and K_{rr}. They even change in sign. Then how do we interpret these constants? At first glance, it would appear that stretching one OH bond weakens the other OH bond if valence forces are assumed but it strengthens the other OH bond if central forces are assumed. This, of course, cannot be true because both potential functions are completely general, and related to one another through the geometry. Therefore, the effect which displacement of one coordinate has on another must be independent of the assumed general potential function. This is readily shown to be true if we convert the *interaction constants to interaction displacement coordinates*. The interaction coordinates are the same for all three force fields of Table 2.11-2. This is apparent from the fact that the interaction coordinate $(l)_k = C_{lk}/C_{kk}$ and the compliance constants C_{ij} are invariant to the choice of force field. Compliance constants are discussed in Sec. 1.13. The compliance constants and interaction coordinates are given in Table 2.11-3. The fact that they are invariant to the choice of coordinates defining

TABLE 2.11-3

COMPLIANCE CONSTANTS AND INTERACTION COORDINATES OF H_2O

Compliance Constants[a,b]		Interaction Coordinates[a]		
C_r	0.119	$(r_2)_{r_1}$	0.020	Å Å^{-1}
$C_{rr'}$	0.00242	$(\alpha)_{r_1}$	$-0.31/r_0$	rad Å^{-1}
$C_\alpha r_0^2$	1.338	$(R_1)_{r_1}$	0.62	Å Å^{-1}
$C_{r\alpha} r_0$	-0.0369	$(r)_\alpha$	$-0.028\ r_0$	Å rad^{-1}
C_R	0.582	$(R)_\alpha$	$0.59\ r_0$	Å rad^{-1}
C_{rR}	0.0737	$(r)_R$	0.13	Å Å^{-1}
$C_{R\alpha} r_0$	0.761	$(\alpha)_R$	$1.37/r_0$	rad Å^{-1}

[a] These values are the same for all three force fields. Units are Angstrom per millidyne for compliance constants and as given for interaction coordinates. $r_0 = 0.957$ Å.

[b] Because of the redundancy among the coordinates (cosine law), there are relations among the compliance constants, only four of which are independent. Thus, $C_{rR} = -\gamma(C_r + C_{rr'}) + \beta C_{r\alpha}$, $C_{R\alpha} = -2\gamma C_{r\alpha} + \beta C_\alpha$, and $C_R = 2\gamma^2(C_r + C_{rr'}) - 4\gamma\beta C_{r\alpha} + \beta^2 C_\alpha$. $\gamma = (\cos\alpha - 1)r_0/R_0$, $\beta = (\sin\alpha)r_0^2/R_0$.

the force field is a strong argument in favor of compliance constants over force constants as a measure of bond strengths and interactions.

The interaction coordinates are the most appropriate potential constants to correlate with the bonding properties. As estimated from the negative value for F_{rr} in the General Quadratic Valence Force Field, the interaction coordinate $(r_2)_{r_1}$ is indeed positive. However, it should be noted that from the Central Force Field of Table 2.11-3 from the positive value of K_{rr} we would predict a negative value for $(r_2)_{r_1}$.

As discussed in Sec. 1.10, the meaning of these interaction coordinates is as follows: (a) If r_1 is stretched by a small amount, *for energy minimization* r_2 would stretch by 2% of the stretch of r_1; (b) if r_1 is stretched by a small amount, *for energy minimization* the angle α would decrease by 0.3 rad Å^{-1} of stretch of r_1; and (c) if the angle α is increased by a small amount, the OH bonds would contract by 0.028 Å rad^{-1} of increase in α *for minimum energy*.

The arguments of Heath and Linnett (*39*) to explain the signs of the interaction constants seem preferable to those of (*38*) and are repeated here as applied to the interaction coordinates. Their argument is essentially that the tetrahedral (sp^3) radius of oxygen is considerably shorter than the divalent radius (p^2), in agreement with the view that sp^3 hybridized bonds are stronger than p-bonds (*40*). This means that the lengthening of either of the OH bonds will lead to an electronic configuration involving more p-bond and less sp^3-bond than for the equilibrium configuration. This increase in p-bond contribution will mean that the whole structure will become more p-bonded which will mean that the equilibrium length of the other bond will increase and the angle will decrease. This would explain the fact that $(r_2)_{r_1}$ is positive and $(\alpha)_{r_1}$ is negative. It must be kept in mind, however, that this argument is *ex post facto* and has no good theoretical basis as yet.

Many other symmetrical bent triatomic molecules have been studied. Generally the harmonic frequencies are much more uncertain than for water and the force constants are less reliable mainly because of the lack of data on isotopic species and the lack of sensitivity of force constant solutions to isotope substitution of heavier atoms. A few of these have been treated by Papousek and Pliva (*41*). Force constants for the SO_2 molecule have been determined using centrifugal distortion data along with vibrational frequencies of the normal and ^{18}O species. Similar studies of NO_2 were made using mean amplitudes and vibrational frequencies. These are treated in the next section.

2.12. Potential Constants of SO_2 and NO_2

The most complete potential constant calculations for SO_2 were made by Ottinger (*42*), using a least squares program which treats the vibrational frequencies of isotopic sqecies, the centrifugal distortion constants, and the mean square amplitudes, when available. They are of special interest as he took pains to demonstrate the effect of excluding certain elements of the data. Such a study is useful for giving one a feeling for the accuracy of compliance constants or force constants determined from a limited amount of data. The results are given in Table 2.12-1.

These calculations have several points of interest. First, it should be recognized that to determine the four force constants we must have at least four pieces of data. In Table 2.12-1, the starred quantities were not used as input data. Calculation I is called by Ottinger (*42*) the "most compatible" solution, as it uses all the data. Calculation II shows that

TABLE 2.12-1

VIBRATIONAL DATA FOR THE SO_2 MOLECULE

		Observed[a]	Calculated[b]				
			I	II	III	IV	V
$S^{16}O_2$	ν_1	1151.4	1151.4	1151.3	1151.4	1151.3*	1441.4*
	ν_2	517.8	517.6	517.6	517.8	516.9*	507.8*
	ν_3	1360.5	1360.4	1360.4	1360.5	1360.3*	1359.6*
$S^{16}O^{18}O$	ν_1	1122.0	1122.5	1122.7	1122.4*	1122.0	1328.5*
	ν_2	506.8	507.3	507.2	507.6*	506.8	497.7*
	ν_3	1341.1	1341.1	1341.0	1341.6*	1341.1	1417.6*
$S^{18}O_2$	ν_1	—	1099.0*	1099.3*	1098.7*	1098.2*	1375.5*
	ν_2	—	496.8*	496.6*	497.2*	496.3*	487.4*
	ν_3	1316.0	1316.3	1316.3	1316.9	1316.2*	1316.0
	t_{aaaa}	143.47	143.72	141.77*	145.87	146.51	143.47
	t_{bbbb}	605.84	646.59	657.97*	632.99	633.38	605.84
	t_{aabb}	−227.64	−199.98	−200.86*	−198.59	−199.25	−227.64
	t_{abab}	23.77	26.10	26.10*	26.09	26.09	26.11
	C_r		0.10029 ±0.00007	0.1003 ±0.0001	0.1003 ±0.0004	0.1004 ±0.0003	0.0822 ±0.0144
	C_α		1.2920 ±0.0181	1.2846 ±0.0255	1.2992 ±0.0283	1.3042 ±0.0299	1.3368 ±0.0903
	C_{rr}		0.000098 ±0.000065	0.0001 ±0.0001	0.0002 ±0.0004	0.0002 ±0.0003	−0.0181 ±0.0144
	$C_{r\alpha}$		−0.0378 ±0.0131	−0.0321 ±0.0177	−0.0435 ±0.0218	−0.0441 ±0.0217	−0.0277 ±0.0273
	F_r		10.08	10.08	10.12	10.12	12.93
	F_α		0.79	0.80	0.79	0.79	0.76
	F_{rr}		0.10	0.10	0.13	0.14	2.94
	$F_{r\alpha}$		0.30	0.26	0.34	0.35	0.33

[a] The units are cm^{-1} for frequencies ν_i, millidynes per Angstrom for the force constants F and Angstrom per millidyne for compliance constants C. The $t_{\alpha\beta\gamma\delta}$ are equal to $-32\pi^4 h^{-3} N^{-2} \times 10^{-37} \times (I^0_{\alpha\alpha} I^0_{\beta\beta} I^0_{\gamma\gamma} I^0_{\delta\delta}) \tau_{\alpha\beta\gamma\delta}$ where the moments of inertia $I_{\alpha\alpha}$ etc. have the units amu Å^2 and $\tau_{\alpha\beta\gamma\delta}$ is in megacycles per second.

[b] The starred quantities were not used as input data.

the frequencies alone give a solution close to the "most compatible" solution. Calculations III and IV show that one can do fairly well using all three frequencies of only one molecule along with the four centrifugal distortion constants. Calculation V gives poor results, showing that one really should have at least all the frequencies for one molecule along with the centrifugal distortion data. Ottinger (*42*) says this is because the centrifugal distortion data is subject to large errors (however, see below).

For the SO_2 frequencies, anharmonicity corrections have not been applied as they are unknown. As pointed out in Sect. 2.14, the neglect of anharmonicity is not nearly so serious in force constant calculations from isotopic frequency shifts for nonhydrogen containing triatomic molecules as it is for hydrogen containing molecules. However it is still a factor to be reckoned with and may be responsible for the rather poor agreement for the centrifugal distortion constants. Furthermore, the centrifugal distortion constants are for the ground vibrational state (not extrapolated to equilibrium). It should be noted that calculation I, which uses all the data, gives the smallest standard deviations in the compliance constants, as expected.

Ottinger (*42*) also evaluated the effect of changing the weight of distortion data relative to frequency data in the least squares analysis. He found that for this treatment of SO_2 by changing the weights of the distortion constants by a factor of 10^4 no great change in calculated compliance constants resulted. This is perhaps not surprising because a weight of zero for distortion constants, Calculation II, gives results not greatly different from Calculation I. This does not mean that in general the weights are not important.

A similar study of NO_2 was made using mean square amplitudes along with vibrational frequencies of $^{14}NO_2$ and $^{15}NO_2$. The results are given in Table 2.12-2. As with SO_2, these calculations show that use of the frequencies alone, for both isotopic species, gives compliance constants not significantly different from those determined using all the data. As expected, use of all the data (Calculation I) leads to smaller standard deviations for the C_{ij}. If frequencies of only one isotopic species are used along with mean square amplitudes (Calculations III and IV), we get a better fit with observed O–O amplitude; however, the other results (I and II) gave Σ_{O-O} well within the reported uncertainty. The frequencies calculated but not used in III and IV are off by up to 2 cm^{-1} from the observed values. This is certainly outside the limits of experimental accuracy, indicating that for this molecule isotopic shifts are more useful than mean amplitudes for estimation of compliance constants. Changing the weights of mean amplitudes relative to those of frequencies was found (*42*) to have little effect on the results.

It is interesting to compare the compliance constants and interaction coordinates of these two molecules as done in Table 2.12-3. The negative interaction coordinate, $(XO_2)_{XO_1}$, for NO_2 is to be expected, as stretching one NO bond would increase the pi bonding in the other NO bond. Note that this interaction coordinate is significantly greater than the analogous values for the linear molecules of Table 2.9-5. The zero value for $(XO_2)_{XO_1}$ for SO_2 is, however, rather surprising. We except rather large SO–SO

TABLE 2.12-2

VIBRATIONAL DATA[a] FOR NO_2

		Observed	I	II[b]	III[b]	IV[b]
	ν_1	1617.75	1617.0	1617.0	1617.8	1616.2*
$^{14}NO_2$	ν_2	749.8	750.0	750.0	749.8	750.2*
	ν_3	1319.7	1319.8	1319.8	1319.7	1317.4*
	ν_1	1580.32	1581.1	1581.1	1582.3*	1580.3
$^{15}NO_2$	ν_2	740.15	740.0	740.0	738.7*	740.2
	ν_3	1306.5	1306.4	1306.4	1308.2*	1306.5
$^{14}NO_2$	Σ_{NO}	0.00146 ±0.00019	0.00151	0.00151*	0.00151	0.00151
	Σ_{OO}	0.00221 ±0.00047	0.00230	0.00230*	0.00221	0.00221
	C_r	0.1011	0.1011 ±0.0003	0.1011 ±0.0004	0.1006 ±0.0001	0.1009 ±0.00002
	C_α		0.9368 ±0.0144	0.9367 ±0.0203	0.9748 ±0.0448	0.9701 ±0.0431
	C_{rr}		−0.0181 ±0.0003	−0.0181 ±0.0004	−0.0184 ±0.0001	−0.0184 ±0.00002
	$C_{r\alpha}$		−0.339 ±0.0051	−0.0339 ±0.0072	−0.0477 ±0.0179	−0.0473 ±0.0185

[a] Units are cm^{-1} for frequencies, $Å^2$ for Σ, and Å $mdyn^{-1}$ for compliance constants.

[b] Starred quantities were not used in the least squares estimates of compliance constants.

interaction from the single bond–double bond resonance structures expected. Perhaps it is not reasonable to put much faith in these interaction coordinates as they could change significantly if anharmonic corrections were made on the frequencies.

The bond strengths are about the same for NO_2 and SO_2 as shown by the similarity in C_{XO}; however, the *smaller* value for C_α of NO_2 indicates that the NO_2 bond directional forces are *stronger* than those for SO_2.

TABLE 2.12-3

COMPLIANCE CONSTANTS AND INTERACTION COORDINATES OF SO_2 AND NO_2

	SO_2		NO_2	
C_{XO}	0.1003	Å $mdyn^{-1}$	0.1011	Å $mdyn^{-1}$
C_α	1.2920		0.9368	
$C_{XO,XO}$	0.0001		−0.0181	
$C_{XO,\alpha}$	−0.0378		−0.0339	
$(XO_2)_{XO_1}$	0.001		−0.18	
$(\alpha)_{XO_1}$	$-0.38/r_{XO}$	rad $Å^{-1}$	$-0.34/r_{XO}$	rad $Å^{-1}$
$(XO_1)_\alpha$	$-0.029\ r_{XO}$	Å rad^{-1}	$-0.036\ r_{XO}$	Å rad^{-1}

2.13. The Nitrosyl Halides

The least symmetrical triatomic molecules are bent, with a different type of atom at each end. There are three fundamental vibrational frequencies and six general quadratic force constants. As pointed out in Sect. 1.11, the vibrational frequencies of at least three different isotopic species are necessary for calculation of the general quadratic force constants unless other data, such as centrifugal distortion constants, is available. For ONF, ONCl, and ONBr, we have knowledge of fundamental harmonic vibrational frequencies of four isotopic species as well as centrifugal distortion constants. Least squares estimates of the general quadratic valence force constants of these three molecules have been made (*43–45*) from the vibrational frequency and centrifugal distortion data. The results of these studies are given in Tables 2.13-1, 2.13-2, and 2.13-3.

In the least squares calculations of force constants, each frequency was weighted as $1/\omega^2$. For the analogous calculation of compliance constants, each ω_i is weighted as ω^6. This gives equal weight to all the frequencies. In the compliance constant program, the centrifugal distortion constant were weighted as K/τ^2 or K/D^2 where K is a scale constant to make $\Sigma P_i(\Delta X_i)^2/n_X = \Sigma P_i(\Delta\varphi_i)^2/n_\varphi$, with P_i = weight, $X = \tau$ or D, and $\varphi = 1/\nu^2$. The weighting procedure is discussed in Sect. 2.9. Nibler (*47*) has discussed the weighting problem in detail. He has shown that for certain systems, such as H_2O–D_2O, the weighting system used is not very important when accurate harmonic frequencies are known but that it has a very significant effect when anharmonic frequencies are used. For the harmonic frequencies of the nitrosyl halides, it does not have a large effect.

The different calculations of Tables 2.13-1 and 2.13-2 have several interesting points. First of all, Calculation I gives the results when all the data (frequencies and centrifugal distortion constants) are used as input while Calculation II uses only the frequencies. For ONF, it is seen that the results are very close for the two cases although inclusion of the centrifugal distortion constants decreases the error limits significantly. The weighting of the $\tau_{\alpha\beta\gamma\delta}$ is discussed in (*43a*). For ONCl, Calculation II is significantly different from Calculation I, perhaps suggesting errors in some of the data.

Calculation III for both ONF and ONCl gives a solution quite different from Calculation II. Thus there are at least two potential energy minima which fit the vibrational frequencies. The centrifugal distortion data serve to reject Calculation III; however, without this additional data one

TABLE 2.13-1

	Observed	I[b]	II[b]	III[b]
$^{16}O^{14}N^{19}F$				
ω_1	1876.8	0.0	0.0	0.0
ω_2	775.5	0.2	0.2	−0.2
ω_3	522.9	0.1	0.1	+0.2
$^{16}O^{14}N^{19}F$				
τ_{xxxx}	−0.0996	−0.0007	−0.0009	0.002
τ_{zzzz}	−15.59	−0.58	−0.65	5.5
τ_{xxzz}	0.419	−0.006	−0.020	0.31
τ_{xzxz}	−0.290	−0.015	−0.013	−0.12
$^{16}O^{15}N^{19}F$				
ω_1	1843.9	0.0	0.0	0.0
ω_2	757.9	0.0	0.0	0.1
ω_3	520.4	0.1	0.1	0.0
$^{18}O^{14}N^{19}F$				
ω_1	1827.1	−0.1	−0.1	0.0
ω_2	767.2	0.0	−0.1	−0.1
ω_3	514.1	−0.2	−0.2	−0.1
$^{18}O^{15}N^{19}F$				
ω_1	1793.4	0.1	0.1	0.0
ω_2	749.5	−0.1	−0.1	0.3
ω_3	512.2	0.1	0.1	−0.2
C_{ON}	—	0.0733[c] ± 0.0011[d]	0.0743[c]	—
C_{NF}	—	0.5291 ± 0.0024	0.5268	—
C_{α}	—	0.5529 ± 0.0023	0.5546	—
$C_{ON,NF}$	—	−0.0741 ± 0.0040	−0.0772	—
$C_{ON,\alpha}$	—	−0.00296 ± 0.00146	−0.00227	—
$C_{NF,\alpha}$	—	−0.0511 ± 0.0023	−0.0487	—
F_{ON}	—	15.93[e] ± 0.04[d]	15.92 ± 0.03[d]	15.64 ± 0.12[d]
F_{NF}	—	2.22 ± 0.04	2.26 ± 0.10	2.04 ± 0.06
F_{α}	—	1.83 ± 0.01	1.82 ± 0.02	2.21 ± 0.12
$F_{ON,NF}$	—	2.24 ± 0.13	2.36 ± 0.31	−1.15 ± 0.25
$F_{ON,\alpha}$	—	0.29 ± 0.05	0.27 ± 0.06	2.36 ± 0.38
$F_{NF,\alpha}$	—	0.22 ± 0.01	0.21 ± 0.02	0.16 ± 0.08

[a] Units are cm^{-1} for ω_1 and Hz sec^{-1} for $\tau_{\alpha\beta\gamma\sigma}$.

[b] Listed in these last three columns for $X(\omega$ and $\tau)$ are $X_{obs.} - X_{calc}$. Calculation I uses the vibrational frequencies and the centrifugal distortion constants. Calculations II and III show two different solutions arrived at from only the frequencies.

[c] Units for C are Å $mdyn^{-1}$ for stretching constants, rad^2 Å^{-1} $mdyn^{-1}$ for bending, and rad $mdyn^{-1}$ for stretch–bend interaction.

[d] Error limits are least squares standard deviations.

[e] Units for F are mdyn Å^{-1} for stretch, mdyn Å rad^{-2} for bend, and mdyn rad^{-1} for stretch–bend interaction.

TABLE 2.13-2

VIBRATIONAL DATA AND FORCE CONSTANTS FOR ONCl

	Observed[a]	I[b]	II[b]	III[b]
$^{16}O^{14}N^{35}Cl$	1835.6	−0.1	0.0	0.0
ω_2	603.2	0.0	0.0	0.1
ω_3	336.4	0.0	−0.1	0.0
D_J	6.35	0.19	0.26	−0.05
D_{JK}	−58.5	−0.4	12.1	−43.9
$^{16}O^{15}N^{35}Cl$				
ω_1	1803.6	0.1	0.0	0.0
ω_2	588.8	0.0	0.0	0.1
ω_3	334.3	0.1	0.1	0.1
$^{18}O^{14}N^{35}Cl$				
ω_1	(1786.4)[c]	—	—	—
ω_2	595.5	0.0	0.0	0.0
ω_3	329.3	0.0	0.0	0.0
$^{18}O^{15}N^{35}Cl$				
ω_1	1753.4	0.0	0.0	0.0
ω_2	580.7	−0.1	−0.1	0.0
ω_3	327.3	−0.1	0.0	−0.1
$^{16}O^{14}N^{37}Cl$				
D_J	6.01	0.11	0.17	—
D_{JK}	−57.4	0.4	12.7	—
C_{ON}[d]	—	0.0746 ± 0.0007[e]	0.08145 ± 0.0062	—
C_{NCl}	—	0.9061 ± 0.0032	0.90099 ± 0.0059	—
C_α	—	0.7655 ± 0.0027	0.75754 ± 0.0038	
$C_{ON,NCl}$	—	−0.0906 ± 0.0035	−0.11969 ± 0.0226	—
$C_{ON,\alpha}$	—	0.0028 ± 0.0032	0.01232 ± 0.0065	—
$C_{NCl,\alpha}$	—	−0.07835 ± 0.0007	−0.06332 ± 0.0124	—
F_{ON}[f]	—	15.26 ± 0.03[e]	15.26 ± 0.03	14.89 ± 0.15
F_{NCl}	—	1.27 ± 0.01	1.38 ± 0.11	1.21 ± 0.04
F_α	—	1.32 ± 0.01	1.31 ± 0.01	1.80 ± 0.19
$F_{ON,NCl}$	—	1.53 ± 0.06	2.02 ± 0.39	−0.90 ± 0.21
$F_{ON,\alpha}$	—	0.10 ± 0.06	−0.08 ± 0.13	2.68 ± 0.52
$F_{NCl,\alpha}$	—	0.12 ± 0.01	0.08 ± 0.04	0.00 ± 0.08

[a] Units are cm^{-1} for ω_i and Kcps for D_J and D_{JK}.
[b] Listed in these last three columns for $X(\omega$ and $D)$ are $X_{obs.} - X_{calc}$.
[c] This is a calculated value for all three calculations. It was not observed.
[d] Units for C are Å $mdyn^{-1}$ for stretching constants, rad^2 Å^{-1} for bending constants, and rad $mdyn^{-1}$ for stretch–bend interactions.
[e] Error limits are least squares standard deviations.
[f] Units for F are mdyn Å^{-1} for stretch, mdyn Å rad^{-2} for bend, and mdyn rad^{-1} for stretch–bend.

would have to choose between Calculations II and III on some arbitrary basis, such as the reasonableness of the force constants. The results for ONBr are analogous; that is, if only frequencies are used, two solutions can be found.

The mean square amplitudes were calculated for the various solutions of Tables 2.13-1, 2.13-2, and 2.13-3 and found not to vary significantly

TABLE 2.13-3

VIBRATIONAL DATA AND FORCE CONSTANTS FOR ONBr

$F_{NO} = 15.25 \pm 0.04^a$ mdyn Å^{-1}	$F_{NO,NBr} = 1.47 \pm 0.36$ mdyn Å^{-1}
$F_{NBr} = 1.13 \pm 0.05$ mdyn Å^{-1}	$F_{NO,\alpha} = 0.11 \pm 0.20$ mdyn rad^{-1}
$F_\alpha = 1.13 \pm 0.02$ mdyn Å rad^{-2}	$F_{NBr,\alpha} = 0.10 \pm 0.02$ mdyn rad^{-1}
$C_{NO} = 0.0749 \pm 0.0049^a$ Å mdyn^{-1}	$C_{NO,NBr} = -0.097 \pm 0.028$ Å mdyn^{-1}
$C_{NBr} = 1.014 \pm 0.026$ Å mdyn^{-1}	$C_{NO,\alpha} = 0.001 \pm 0.014$ rad mdyn^{-1}
$C_\alpha = 0.890 \pm 0.022$ rad^2 Å^{-1} mdyn^{-1}	$C_{NBr,\alpha} = -0.076 \pm 0.007$ rad mdyn^{-1}

	ω_1 (cm^{-1})		ω_2 (cm^{-1})		ω_3 (cm^{-1})	
	obs.	calc.	obs.	calc.	obs.	calc.
$^{16}O^{14}NBr$	1832.3	1832.4	548.0	548.0	269.2	269.3
$^{16}O^{15}NBr$	1800.4	1800.3	—	534.4	267.0	266.8
$^{18}O^{14}NBr$	1783.0	1782.9	—	541.0	—	261.7
$^{18}O^{15}NBr$	1749.9	1750.0	527.1	527.1	259.7	259.6
$^{16}O^{14}N^{79}Br$	—	1832.4	—	548.1	—	269.7

	D_J (kHz)		D_{JK} (kHz)	
	obs.	calc.	obs.	calc.
$^{16}O^{14}N^{79}Br$	3.05 ± 0.1	2.84	−54.8 ± 2	−54.6

[a] Error limits are least squares standard deviations.

with force constant solution. Thus mean square amplitudes would not be particularly useful for determination of force constants.

In order to describe the motion, the eigenvectors (relative amplitudes) and potential energy distributions are given for the best solution (Calculation I) for each $^{16}O^{14}NX$ molecule, with X = F or Cl. The results for ONBr are similar (*45*).

Both Tables 2.13-4 and 2.13-5 show that the high frequency for each molecule is predominantly NO stretch. Though the middle and low frequencies are strong mixtures of both ONX bending and NX stretching motions, it is apparent that the middle frequency is more bending and the low frequency more NX stretch. This is somewhat surprising because the

TABLE 2.13-4

EIGENVECTORS FOR ONF AND ONCl

	ONF			ONCl		
ν	1876.8	775.5	522.9	1835.6	603.2	336.4
L_{NO}	0.36	−0.05	−0.02	0.36	−0.03	−0.02
L_{NX}	−0.02	0.27	0.23	−0.05	0.23	0.21
L_{α}	−0.16	−0.37	0.16	−0.11	−0.36	0.10

lowest frequency has a much greater isotope shift for $^{16}O \rightarrow {}^{18}O$ than for $^{14}N \rightarrow {}^{15}N$. Thus, one must be cautious in making frequency assignments from isotope shifts on a qualitative basis.

A comparison of the bonding in ONF, ONCl, and ONBr can be made from the data of Tables 2.13-1, 2.13-2, and 2.13-3. The NO force constant of ONF is the same as that for the free NO molecule (15.9 mdyn Å–see Appendix V). We see that for ONCl and ONBr it is significantly weaker, indicating greater interaction of chlorine with NO than of fluorine with NO in these nitrosyl halides. For all three of these molecules, the nitrogen–halide bond distances are more than 10% longer than a normal single bond, probably because of oxygen–halide repulsion. The N–halide bond must involve a partly occupied antibonding pi orbital of NO and a *p* pi orbital of the halide but it seems that the oxygen–halide interaction must be considered also in describing the electronic structure.

Another interesting comparison is that of the effect one coordinate has on another as described by the interaction constants. The latter are best discussed by referring to the interaction displacement coordinates (Table 2.13-6). Thus, the value of −1.2 for $(NCl)_{NO}$ tells us that if, in ONCl,

TABLE 2.13-5

POTENTIAL ENERGY DISTRIBUTION FOR ONF AND ONCl

	ONF			ONCl		
ν	1876.8	775.5	522.9	1835.6	603.2	336.4
V_{NO}	1.00	0.13	0.04	1.02	0.06	0.06
V_{NX}	0.0	0.47	0.75	0.0	0.32	0.83
V_{α}	0.02	0.69	0.30	0.1	0.81	0.19
$V_{NO,NX}$	0.0	−0.20	−0.17	−0.03	−0.10	−0.16
$V_{NO,\alpha}$	−0.02	0.03	−0.01	0.0	0.01	0.0
$V_{NX,\alpha}$	0.0	−0.11	0.09	0.0	−0.10	0.08

the NO bond is constrained to a small displacement from its equilibrium length, the electrons in the molecule will rearrange themselves and tend to cause a change of the NCl bond by −120% of the change of NO. In other words, if the NO bond is constrained to a small displacement of +1, the equilibrium structure for the molecule would have the NCl bond shortened by −1.2. This does not tell us what motion occurs during a vibration but does help tell us how difficult the vibrational motion

TABLE 2.13-6

INTERACTION DISPLACEMENT COORDINATES[a] OF ONCl

	ONF	ONCl	ONBr
$(NX)_{NO}$	-1.01 ± 0.04	-1.21 ± 0.04	-1.29 ± 0.29
$(\alpha)_{NO}$	-0.04 ± 0.02 rad Å^{-1}	$+0.04 \pm 0.04$	0.01 ± 0.19
$(OX)_{NO}$	-0.2	-0.4	-0.5
$(NO)_{NX}$	-0.140 ± 0.008	-0.1 ± 0.004	-0.096 ± 0.025
$(\alpha)_{NX}$	-0.097 ± 0.004 rad Å^{-1}	-0.086 ± 0.001	-0.075 ± 0.005
$(OX)_{NX}$	$+0.7$	$+0.8$	$+0.8$
$(NO)_{\alpha}$	-0.005 ± 0.003 Å rad^{-1}	$+0.004 \pm 0.004$	0.001 ± 0.016
$(NX)_{\alpha}$	-0.092 ± 0.004 Å rad^{-1}	-0.102 ± 0.001	-0.086 ± 0.009

[a] The interaction displacement coordinate $(l)_k$ is the displacement of coordinate l which would arise for *minimization of energy* if coordinate k is constrained to a small unit positive displacement.

is. It is also an indication of the change in electronic structure accompanying a displacement of a particular coordinate. The coordinates in Table 2.13-6 are quite similar for all three molecules. Therefore, we are sure that the bonding is quite similar. The NO,NX interaction is quite large, indicating considerable electronic rearrangement if one bond is constrained in a displaced position.

The interaction coordinates can be estimated from molecular orbital theory by calculating the minimum energy structure with one coordinate constrained in a displaced position. Such a calculation will be very interesting for comparison with the interaction coordinates derived from the compliance constants $[(l)_k = C_{lk}/C_{kk}]$. In this respect, the compliance constants are more interesting than the force constants. Furthermore they are more unique; they do not depend on our choice of important internal coordinates as force constants do [see (*43b*)].

2.14. Use of Anharmonic Frequencies

We have used harmonic frequencies for the force constant calculations presented here when possible. Many times anharmonicity corrections are not available so one uses the anharmonic (observed) frequencies for force constant calculations. It is important to know how force constants calculated from the anharmonic frequencies may differ from the true harmonic values. Nibler (*47*) has made such an investigation for several triatomic molecules. His results for H_2O are listed in Table 2.14-1.

TABLE 2.14-1

Force Constants[a] for H_2O

	Harmonic	Anharmonic
F_r	8.45	7.57
F_{rr}	−0.10	−0.27
$\sqrt{2}F_{r\alpha}/r_0$	0.33	−0.63
F_α/r_0^2	0.76	0.80

[a] Units are mdyn Å^{-1}.

These values were calculated from the frequencies for H_2O, D_2O, and HDO. It is apparent that the interaction constants are greatly in error for this case if the observed frequencies are used. Furthermore he (*47*) has shown that the weighting scheme is very important when anharmonic frequencies are used; quite different solutions are obtained for different weighting schemes. The best choice of weighting is not yet established and certainly varies among individual cases. For harmonic frequencies, the weighting scheme has a much smaller effect on the results.

Similar calculations have been made by the author for HCN. Force constant solutions have been calculated from the frequencies of HCN and DCN using the secular equation (III-6) for both molecules (three force constants from four frequencies).

The results are given in Table 2.14-2.

The frequencies were weighted equally ($1/\omega_i^2$ used as the weighting element).

In spite of the rather large anharmonicity corrections, the anharmonic force constants are merely scaled down from the harmonic force constants, though the frequency fit is rather poor for the anharmonic treatment. The

TABLE 2.14-2

VIBRATIONAL FREQUENCIES AND FORCE CONSTANTS FOR HCN AND DCN

		Harmonic	Anharmonic
		(cm^{-1})	(cm^{-1})
ω_1	HCN	2132.0	2096.7
	DCN	1953.5	1925.3
ω_3	HCN	3438.5	3311.5
	DCN	2703.6	2630.3
		(mdyn Å^{-1})	(mdyn Å^{-1})
F_{HC}		6.23 ± 0.00	5.79 ± 0.07
F_{CN}		18.77 ± 0.00	18.32 ± 0.26
$F_{HC,CN}$		-0.22 ± 0.00	-0.17 ± 0.12

uncertainty in the force constants is also much greater for the anharmonic treatment.

For the thiocyanate ion, similar calculations have been made and are given in Table 2.14-3. For this case, the limits of error are about the same for harmonic and anharmonic force constants. For hydrogen containing molecules, the anharmonicity corrections are much greater for H than for D so whether they are included or not makes considerable difference. For nonhydrogen containing molecules, such as NCS^-, the

TABLE 2.14-3

VIBRATIONAL FREQUENCIES AND FORCE CONSTANTS OF NCS^-

		Harmonic	Anharmonic
		(cm^{-1})	(cm^{-1})
$^{14}N^{12}C^{32}S$	ω_1	2085.8	2053.0
	ω_3	757.0	748.7
$^{14}N^{13}C^{32}S$	ω_1	2037.3	2006.0
	ω_3	750.8	742.6
$^{14}N^{12}C^{34}S$	ω_1	2084.8	2052.0
	ω_3	747.3	739.2
$^{14}N^{13}C^{34}S$	ω_1	2036.3	2005.0
	ω_3	741.6	733.1
F_{CN}		16.02 ± 0.15	15.80 ± 0.17
F_{CS}		5.06 ± 0.03	4.90 ± 0.03
$F_{CN,CS}$		0.96 ± 0.12	1.16 ± 0.14

TABLE 2.14-4

HARMONIC AND ANHARMONIC FORCE CONSTANTS[a] FOR ONF

	With $\tau_{\alpha\beta\gamma\delta}$		Without $\tau_{\alpha\beta\gamma\delta}$	
	Harmonic	Anharmonic	Harmonic	Anharmonic
F_{NO}	15.93	15.27	15.92	15.21
F_{NF}	2.22	2.28	2.26	2.81
F_{α}	1.83	1.79	1.82	1.73
$F_{NO,NF}$	2.24	2.47	2.36	3.61
$F_{NO,\alpha}$	0.29	0.15	0.27	0.13
$F_{NF,\alpha}$	0.22	0.19	0.21	0.12

[a] Units are mdyn^{-1}.

anharmonicity corrections for the isotopic species are not greatly different so they act primarily as a scale factor for these simple cases of 2 × 2 secular equations.

In order to investigate a more complicated case, we have calculated the force constants of ONF and ONCl from the anharmonic (observed) frequencies. The results are given in Tables 2.14-4 and 2.14-5. Calculations were made with and without the centrifugal distortion constants used as input. From these tables, it is apparent that if the centrifugal distortion constants are used, the force constants for the anharmonic frequencies are not greatly different from the harmonic values. However, if the frequencies alone are used, the anharmonic values for F_{NX} and $F_{NO,NX}$ vary widely from the harmonic values. It is only the high frequency (NO stretch) which has an anharmonic correction of more than a few wave numbers.

TABLE 2.14-5

HARMONIC AND ANHARMONIC FORCE CONSTANTS[a] FOR ONCl

	With D_J and D_{JK}		Without D_J and D_{JK}	
	Harmonic	Anharmonic	Harmonic	Anharmonic
F_{NO}	15.26	14.45	15.26	14.47
F_{NCl}	1.27	1.24	1.38	2.46
F_{α}	1.32	1.30	1.31	1.22
$F_{NO,NCl}$	1.53	1.47	2.02	4.33
$F_{NO,\alpha}$	0.10	−0.24	−0.08	−0.50
$F_{NCl,\alpha}$	0.12	0.08	0.08	−0.18

[a] Units are mdyn Å^{-1}.

This shows that for systems of this complexity (3 × 3 or greater), for calculation of force constants from isotopic frequencies only, it is desirable to have accurate values corrected for anharmonicity, if the correction is more than a few wave numbers.

REFERENCES

1. E. R. Cohen and J. W. M. Dumond, *Rev. Mod. Phys.*, **37,** 590 (1965).
2. D. H. Rank, A. G. St. Pierre, and T. A. Wiggins, *J. Mol. Spectry.*, **18,** 418 (1965).
3. G. Herzberg, *Spectra of Diatomic Molecules* Van Nostrand, Princeton, New Jersey, 1950, pp. 501–581.
4. D. H. Rank, D. P. Eastman, B. S. Rao, and T. A. Wiggins, *J. Opt. Soc. Am.*, **52,** 1 (1962).
5. C. A. Coulson, *Valence* (2nd ed.), Oxford, New York, 1961, p. 91ff.
6. C. P. Courtoy, *Ann. Soc. Sci. Bruxelles*, **Ser. 1, 73,** 5 (1959).
7. G. Amat and M. Pimbert, *J. Mol. Spectry.*, **16,** 278 (1965).
8. H. R. Gordon and T. K. McCubbin, Jr., *J. Mol. Spectry.*, **19,** 137 (1966).
9. S. F. A. Kettle, *Spectrochim. Acta*, **22,** 1388 (1966).
10. J. Pliva, *Collection Czech. Chem. Commun.*, **23,** 1846 (1958).
11. M. H. Pariseau, I. Suzuki, and J. Overend, *J. Chem. Phys.*, **42,** 2335 (1965).
12. K. Kuchitsu and Y. Morino, *Bull. Chem. Soc. Japan*, **38,** 5 (1965).
13. B. P. Stoicheff, *Can. J. Phys.*, **36,** 218 (1958).
14. D. Agar, E. K. Plyer, and E. D. Tidwell, *J. Res. Natl. Bur. Std.*, **66A,** 259 (1962).
15. T. Wentink, Jr., *J. Chem. Phys.*, **29,** 188 (1958); *ibid.*, **30,** 105 (1959).
16. L. H. Jones, *Inorg. Chem.*, **6,** 429 (1967).
17. H. H. Claassen, G. L. Goodman, J. G. Malm, and F. Schreiner, *J. Chem. Phys.*, **42,** 1229 (1965).
18. P. A. Agron, G. M. Begun, H. A. Levy, A. A. Mason, C. G. Jones, and D. F. Smith, *Science*, **139,** 24 (1962).
19. K. O. Christe, W. Sawodny, and J. P. Guertin, *Inorg. Chem.*, **6,** 1159 (1967).
19*a*. J. C. Evans and G. Y.-S. Lo, *J. Chem. Phys.*, **44,** 4356 (1966).
19*b*. G. C. Hayward and P. J. Hendra, *Spectrochim. Acta*, **23A,** 2309 (1967).
20. R. E. Rundle, *J. Am. Chem. Soc.*, **85,** 112 (1963); R. J. Hach and R. E. Rundle, *ibid.*, **73,** 4321 (1951).
21. G. C. Pimentel and R. D. Spratley, *J. Am. Chem. Soc.*, **85,** 826 (1963); G. C. Pimentel, *J. Chem. Phys.* **19,** 446 (1951).
22. C. A. Coulson, *J. Chem. Phys.*, **44,** 468 (1966).
23. D. F. Smith, in *Noble Gas Compounds* (H. H. Hyman, ed.), Univ. Chicago Press, Chicago, Illinois, 1963), p. 295.
24. H. Papazian, *J. Chem. Phys.*, **34,** 1614 (1961).
25. J. I. Bryant and G. C. Turrell, *J. Chem. Phys.*, **37,** 1069 (1962).
26. J. I. Bryant, *J. Chem. Phys.*, **40,** 3195 (1964).
27. L. H. Jones, unpublished work.
28. J. Pliva, *J. Mol. Spectry.*, **27,** 461 (1968).
28*a*. J. Pliva, *J. Mol. Spectry.*, **12,** 360 (1961).
29. R. P. Grosso and T. K. McCubbin, Jr., *J. Mol. Spectry.*, **13,** 240 (1964).

30. J. L. Griggs, Jr., K. N. Rao, L. H. Jones, and R. M. Potter, *J. Mol. Spectr.* **18,** 212 (1965).
31. A. Maki and J. C. Decius, *J. Chem. Phys.*, **31,** 772 (1959).
32. L. H. Jones, *J. Chem. Phys.*, **25,** 1069 (1956).
33. D. H. Rank, G. Skorinko, D. P. Eastman, and T. A. Wiggins, *J. Opt. Soc. Am.*, **50,** 421 (1960).
34. H. C. Allen, Jr., E. D. Tidwell, and E. K. Plyer, *J. Chem. Phys.*, **25,** 302 (1956).
35. I. Suzuki, M. A. Pariseau, and J. Overend, *J. Chem. Phys.*, **44,** 3561 (1966).
36. W. S. Benedict, N. Gailar, and E. K. Plyer, *J. Chem. Phys.*, **24,** 1139 (1956).
37. K. Kuchitsu and L. S. Bartell, *J. Chem. Phys.*, **36,** 2460 (1962).
38. C. A. Coulson, J. Duchesne, and C. Manneback, *Victor Henri Memorial Volume*, Desoer, Liege, 1947, p. 33.
39. D. F. Heath and J. W. Linnett, *Trans. Faraday Soc.*, **44,** 556 (1948).
40. L. Pauling and Sherman, *J. Am. Chem. Soc.*, **59,** 1450 (1937).
41. D. Papousek and J. Pliva, *Collection Czech. Chem. Commun.*, **29,** 1973 (1964).
42. R. S. Ottinger, Ph. D. Thesis, Oregon State University, 1966.
42*a*. R. R. Ryan, unpublished work.
43*a*. R. R. Ryan and L. H. Jones, *J. Chem. Phys.*, **50,** 1492 (1969); *b*. L. H. Jones, L. B. Asprey, and R. R. Ryan, *J. Chem. Phys.*, **47,** 3371 (1967).
44. L. H. Jones, R. R. Ryan, and L. B. Asprey, *J. Chem. Phys.*, **49,** 581 (1968).
45. J. Laane, L. H. Jones, R. R. Ryan, and L. B. Asprey, *J. Mol. Spectry.*, **30,** 485 (1969).
46. S. F. A. Kettle, *Spectrochim. Acta*, **22,** 1388 (1966).
47. J. W. Nibler, Ph.D. Thesis, University of California at Berkeley, Berkeley, California, 1966.

Chapter 3

Force Constants of Halide Complexes

The vibrational spectra of a large number of metal halide and nonmetal halide complexes have been studied. Though the secular equations are comparatively simple, there are still only a few cases for which general quadratic force constants have been determined unequivocally. In such cases the solution has been made possible by the inclusion of isotopic frequency shifts or Coriolis coupling constants or both and, in a few cases, centrifugal distortion data. In many cases where sufficient information was lacking for a general treatment, somewhat restricted force fields have been applied. Many of the results are presented and discussed in this chapter.

3.1. The Boron Trihalides

There have been many investigations of the planar trihalides of boron of D_{3h} symmetry; e.g., BF_3 (*1–9*), BCl_3 (*1–4, 6*), BBr_3 (1 -*4, 10*), and BI_3 (*3, 4, 10*). There are four fundamental vibrations; ν_1 of A_1' symmetry, ν_2 of A_2'' symmetry, and ν_3 and ν_4 of E' symmetry. The relations among frequencies and force constants and the meanings of the force constants are discussed in Appendix VI. It is shown that there are five force constants to determine from four frequencies for a general quadratic valence force field; F_r, F_{rr}, $H_\alpha = F_\alpha - F'_{\alpha\alpha}$, $H_{r\alpha} = F''_{r\alpha} - F'_{r\alpha}$, and $\bar{F}_\pi$. The subscripts given refer to the internal coordinates; r is a bond stretch, α is the in-plane XMX angle, and π is the angle a given MX bond makes with the plane of the other two MX bonds. The constant $\bar{F}_\pi$ is completely determined by the frequency ν_2; however, the other four force constants determine only three vibrational frequencies. One must have other information to obtain a complete solution.

We shall first discuss the BF_3 molecule which has been the subject of many investigations. Lindeman and Wilson (*1*) have used the frequencies of both isotopic species, $^{10}BF_3$ and $^{11}BF_3$, which give sufficient information for calculation of the five force constants. For this method to be accurate, one should use the harmonic frequencies (corrected from the observed values for anharmonicity). Pistorius (*2*) attempted to do this; however, some of his anharmonic corrections are in the wrong direction and are probably incorrect. Duncan (*9*) estimated the Coriolis coupling constant for the E bending mode, ν_4, of $^{11}BF_3$, to be -0.81 ± 0.02. This estimate was from the band contours as measured by Gage and Barker (*11*) and by Dreska et al. (*8*). This ζ value is sensitive to the force constants and thus serves to fix the solution. However, Levin and Abramowitz (*4*) remeasured ν_4 of $^{11}BF_3$ and estimated that $\zeta_4 = -0.66 \pm 0.03$, yielding a considerably different force constant solution. These results are in significant disagreement and perhaps it will take high resolution studies of ν_4 to choose between them. However, the treatment by Duncan was much more comprehensive and is to be preferred at present.

Konaka et al. (*6*) have determined the mean vibrational amplitudes from electron diffraction studies of BF_3. The mean amplitudes give additional data for calculation of force constants; however, their results show that the amplitudes are not very sensitive to the force constants in this case and so are not useful for solving the BF_3 problem. McKean (*5*) has shown that the isotopic shift of ν_4, from $^{10}BF_3$ to $^{11}BF_3$, is fairly sensitive to the force constant solution. Perhaps when this is known more accurately, it will yield the best force constant solution for BF_3.

The most widely accepted observed frequency values are those given in Table 3.1-1. From the observed combination bands of Vanderryn (*12*) and Jones (*13*), most of the anharmonic corrections can be estimated. These estimates are given in the table. In order to satisfy the harmonic frequencies we calculate that $F_{34} = H_{r\alpha} = -0.4$ mdyn rad^{-1}. This agrees with Duncan's (*9*) ζ value but not that of Levin and Abramowitz (*4*). The harmonic force constants are given in Table 3.1-1. To fit the ζ value of Levin and Abramowitz (*4*) would require that $F_{34} \approx -1.0$ rather than -0.4.

Thus the ζ value is quite a sensitive measure of the force constants. Further high resolution studies of the ν_4 bands of enriched $^{10}BF_3$ and $^{11}BF_3$ are desirable for better determination of the force field.

It is of interest to examine the significance of the interaction constants of BF_3. Thus, use of Eqs. (VI-5) and (VI-6) of Appendix VI with the

TABLE 3.1-1

VIBRATIONAL FREQUENCIES AND FORCE CONSTANTS[a] OF BF_3

	$^{10}BF_3$		$^{11}BF_3$	
	Observed	Harmonic	Observed	Harmonic
ν_1	888	903.6	888	903.6
ν_2	719.5	—	691.3	—
ν_3	1504.7	1512.2	1454.0	1460.2
ν_4	482.0	488.7	480.4	487.0

$X_{11} = -1.8, -1.8$ | $X_{33} = -0.7, -0.6$
$X_{13} = -6.7, -6.4$ | $X_{34} = -1.9, -1.8$
$X_{14} = -5.4, -5.4$ | $X_{44} = -0.7, -0.7$
$F_{11} = 9.15$ mdyn Å^{-1} | $F_r = 7.4$ mdyn Å^{-1}
$F_{33} = 6.5$ mdyn Å^{-1} | $F_{rr} = 0.9$ mdyn Å^{-1}
$F_{34} = -0.4$ mdyn rad^{-1} | $H_{r\alpha} = F''_{r\alpha} - F'_{r\alpha} = F_{34}$
$F_{44} = 0.9$ mdyn Å rad^{-2} | $H_{\alpha} = F_{\alpha} - F'_{\alpha\alpha} = F_{44}$
$F_{22} = \bar{F}_{\pi} = 0.29$ mdyn Å rad^{-2}

[a] All values of ν are in cm^{-1}. The X figures are estimated anharmonic corrections in cm^{-1}. [See Eq. (1.3-1).] The first entry is for $^{10}BF_3$, the second for $^{11}BF_3$.

force constants of Table 3.1-1 leads to the following interaction displacement coordinates:

$$(r_2)_{r_1} = -0.1;$$
$$(\alpha_{12})_{r_1} = -0.2 \quad \text{rad Å}^{-1};$$
$$(\alpha_{23})_{r_1} = +0.4 \quad \text{rad Å}^{-1};$$
$$(r_3)_{\alpha_{12}} = +0.06 \quad \text{Å rad}^{-1};$$
$$(r_1)_{\alpha_{12}} = -0.03 \quad \text{Å rad}^{-1}$$

The meaning of these quantities, as explained previously, follows from the meaning of $(r_2)_{r_1}$ as the displacement of r_2 which leads to minimization of energy after small unit positive displacement of r_1. Thus, if we stretch r_1 a small amount, r_2 would contract by about 10% of that amount for minimization of energy. There are several possible explanations for such an effect; for example, stretching one fluorine away from the boron atom increases the positive charge on the boron and makes it able to pull tighter on the other fluorine atoms.

Similarly we see that stretching of a BF bond leads to contraction of the FBF angles which contain that bond. This also could arise from several effects such as F–F repulsion, or changes in *sp* hybridization accompanying the distortion.

In an isolated case such as this, the information on change in electronic structure accompanying distortion is not specific. However, when such

TABLE 3.1-2

FUNDAMENTAL FREQUENCIES[a] FOR BX_3

	$^{10}BCl_3$	$^{11}BCl_3$	$^{10}BBr_3$	$^{11}BBr_3$	$^{10}BI_3$	$^{11}BI_3$
ν_1	471	471	278	278	190	190
ν_2	480	460	395	375	352	336
ν_3	995	956	856	820	737	704
ν_4	244[b]	243[b]	150	150	100	100

[a] Values are in cm^{-1}.
[b] Reference (*4*) gives 255 cm^{-1}.

information is available on a large number of molecules, we can hope to correlate the interaction displacement coordinates with specific changes in electronic configuration accompanying distortion of one coordinate.

The other trihalides of boron are less well understood. The consensus of infrared and Raman results (*1–4*, *10*) yields the assignments of Table 3.1-2. The isotopic shifts and ζ values for the heavier boron trihalides are less well known than for BF_3 so the force constants are even more uncertain. However, a crude calculation indicates that the primary force constants decrease with increased mass of the halide, as expected.

A Urey–Bradley treatment of BF_3 gives agreement (*14*) with the isotope shifts; however, at present its validity is not established (see Appendix VII and Sect. 1.12).

3.2. METAL HEXAHALIDES OF O_h SYMMETRY

A large number of metal hexafluorides have been prepared and the vibrational spectra of many of them has been studied in the vapor phase. Force constants have been calculated; though, here again, for a general quadratic force field there are seven force constants to calculate from six frequencies. The forms of the normal vibrations, suitable symmetry coordinates, and the F and G matrices are given in Appendix VIII.

The six vibrations for MX_6 are divided in the representations $A_{1g} + E_g + 2F_{1u} + F_{2g} + F_{2u}$. Three are stretching vibrations ($A_{1g} + E_g + F_{1u}$); the other three are bending vibrations. All the g frequencies are Raman-active only; the F_{1u} frequencies are infrared-active only; the F_{2u} frequency is inactive and can only be determined from combination frequencies. The force constant–frequency relations are given below:

$$A_{1g} \quad \lambda_1 = F_{11}\mu_X \tag{3.2-1}$$

$$E_g \quad \lambda_2 = F_{22}\mu_X \tag{3.2-2}$$

$$F_{1u} \quad \lambda_3 + \lambda_4 = F_{33}(2\mu_M + \mu_X) + F_{44}\rho^2(8\mu_M + 2\mu_X) - 8F_{34}\rho\mu_M \tag{3.2-3}$$

$$\lambda_3\lambda_4 = (F_{33}F_{44} - F_{34}^2)\rho^2(12\mu_M\mu_X + 2\mu_X^2)$$

$$F_{2g} \quad \lambda_5 = 4F_{55}\rho^2\mu_X \tag{3.2-4}$$

$$F_{2u} \quad \lambda_6 = 2F_{66}\rho^2\mu_X \tag{3.2-5}$$

In the above equations, $\lambda_i = 4\pi^2\nu_i^2$; $\mu_M = 1/M_M$; $\mu_X = 1/M_X$; $\rho = 1/r$ and

$$F_{11} = F_r + 4F''_{rr} + F'_{rr} \tag{3.2-6}$$

$$F_{22} = F_r - 2F''_{rr} + F'_{rr} \tag{3.2-7}$$

$$F_{33} = F_r - F'_{rr} \tag{3.2-8}$$

$$F_{44} = (F_\alpha - F''_{\alpha\alpha}) + 2(\bar{F}'_{\alpha\alpha} - F'''_{\alpha\alpha}) \tag{3.2-9}$$

$$F_{34} = 2(F'_{r\alpha} - F''_{r\alpha}) = 2H'_{r\alpha} \tag{3.2-10}$$

$$F_{55} = 2(F_\alpha - F''''_{\alpha\alpha}) - (F_\alpha - F''_{\alpha\alpha}) \tag{3.2-11}$$

$$F_{66} = (F_\alpha - F''_{\alpha\alpha}) - 2(\bar{F}'_{\alpha\alpha} - F'''_{\alpha\alpha}) \tag{3.2-12}$$

There are seven valence force constants to determine: F_r, F''_{rr}, F'_{rr}, $F_\alpha - F''_{\alpha\alpha}$, $F_\alpha - F''''_{\alpha\alpha}$, $\bar{F}'_{\alpha\alpha} - F'''_{\alpha\alpha}$, and $H'_{r\alpha}$. The force constants are defined in Appendix VIII.

The various bending constants occur only in combinations because of the interdependency among the angles (one angle cannot be distorted without affecting at least one other angle). Thus, we can determine $F_\alpha - F''_{\alpha\alpha}$ but not F_α or $F''_{\alpha\alpha}$ by themselves. In fact, by themselves they have no physical meaning; however, the corresponding compliance constants do (see Appendix VIII).

The only valence force constant which can be determined from only the frequency data of the normal species is F''_{rr}, the constant for interaction of bonds at 90° to each other. Thus from Eqs. (3.2-1), (3.2-2), (3.2-6), and (3.2-7),

$$F''_{rr} = (\lambda_1 - \lambda_2)/6\mu_X \tag{3.2-13}$$

Claassen (*15*) has calculated this quantity for nine different hexafluorides (Mo, W, Re, Os, Ir, Pt, U, Np, and Pu). He found fairly appreciable

TABLE 3.2-1

General Quadratic Valence Force Constants[a] of Some MF_6 Molecules

Molecule	F_r	F^c_{rr}	F^t_{rr}	$\rho(F'_{r\alpha} - F''_{r\alpha})$[b]	$\rho^2(F_\alpha - F''_{\alpha\alpha})$	$\rho^2(\bar{F}'_{\alpha\alpha} - F'''_{\alpha\alpha})$	$\rho^2(F_\alpha - F''''_{\alpha\alpha})$
RhF_6	4.21	0.10	−0.09	0.05	0.25	0.02	0.22
WF_6	5.19	0.26	0.41	0.10	0.19	0.05	0.23
IrF_6	4.90	0.14	0.05	0.08	0.28	0.02	0.23
PtF_6	4.50	0.13	−0.22	0.08	0.29	0.02	0.23
UF_6	3.73	0.30	0.06	−0.06	0.14	0.01	0.12
NpF_6	3.67	0.26	−0.02	−0.03	0.17	0.01	0.14
PuF_6	3.56	0.23	−0.05	−0.03	0.18	0.01	0.15
SF_6[c]	5.01	0.34	0.26	0.37	0.89	0.11	0.83
SeF_6[c]	5.01	0.12	0.13	0.22	0.51	0.06	0.49
TeF_6[c]	5.10	0.07	0.12	0.12	0.31	0.05	0.29

[a] Units of all force constants given are mdyn Å^{-1}.
[b] ρ is the reciprocal of the *M*–F distance.
[c] The data for these molecules comes in part from (*77*).

interaction, F''_{rr}, varying from 0.13 to 0.39 mdyn Å^{-1} in this group of compounds. There appears to be no systematic trend except that within a given group of the periodic table the interaction constant, F''_{rr}, increases with the primary stretching force constant, F_r. More recently Kim et al. (*16*) have calculated complete quadratic force fields for ten hexafluorides using Coriolis constants derived from band contours (*16*, *17*) along with the vibrational frequencies. Unique force constant solutions are obtained in this manner; the general quadratic valence force constants calculated from the data of (*16*) are given in Table 3.2-1. It is difficult to put error limits on these numbers as anharmonicity corrections are not known and the uncertainties in zeta values derived from band contours may be rather large.

The force constants listed in Table 3.2-1 are defined by the general quadratic valence force potential function given as Eq. (VIII-1) in Appendix

VIII. Kim et al. (*16*) also calculated force constants for these molecules in a Urey–Bradley force field (UBFF) and an orbital valency force field (OVFF). The respective potential functions are given below:

$$2V = K\sum_{i}^{6}(\Delta r_{MF})^2 + \left\{\begin{matrix} r_0^2 H \sum_{ij}^{12}(\Delta\alpha_{ij})^2 \\ \\ r_0^2 D \sum_{i}^{6}(\Delta\beta_i)^2 \end{matrix}\right\} + F\sum_{ij}^{12}(\Delta R_{\mathrm{F-F}})^2$$
$$+ 2R_0F'\sum_{ij}^{12}\Delta R_{\mathrm{F-F}} + 2r_0f'\sum_{i}^{6}\Delta r_{MF}$$

The upper term in the brackets is for the UBFF; the lower term is for the OVFF. Thus, the only difference is in the type of angle bending considered. For the UBFF, $\Delta\alpha_{ij}$ is a change in an F*M*F angle; for OVFF, $\Delta\beta_i$ is a change in angle the *i*th *M*F bond makes with the axis of the *i*th octahedral sigma bond orbital. The quantity f' is related to F' so there are only four force constants to determine from six frequencies for the UBFF and for the OVFF. Kim et al. (*16*) determined force constant solutions to "best fit" the six frequencies and found that from the force constant solutions the calculated frequencies agreed much better for the OVFF than for the UBFF. Thus, their results offer some evidence in favor of an Orbital Valence Force Field; however, the calculated and observed frequencies still differ by up to 40 cm^{-1} for the OVFF. Therefore, a unique set of general quadratic force constants are to be preferred and the orbital valency force field cannot be considered as proven for these molecules.

The force constants of Table 3.2-1 show some interesting points. The stretching constants, F_r, are definitely weaker for the actinide metals than for the others. Even though the bonds become shorter in the order $R_{\mathrm{UF}} > R_{\mathrm{NpF}} > R_{\mathrm{PuF}}$, they become weaker in the same order. This is the same phenomenon observed for the MO bonds in the actinyl ions in $NaMO_2(Ac)_3$ (*18*). The explanation given (*18*) was that interaction with the extra valence shell electrons weakens the bonds. The primary bending constants show a definite reciprocal relation to the MF bond lengths; this strongly suggests that the F–F repulsion is important for the molecules with shorter bonds (such as SeF_6 and SF_6).

The interaction constants can best be discussed by converting to interaction displacement coordinates using relations (VIII-3) through (VIII-8) in Appendix VIII. As discussed in Sect. 1.10, the interaction displacement coordinate $(\alpha)_\beta$ is defined as follows. If coordinate β is constrained to a unit positive displacement, the electrons in the molecule will rearrange slightly. To minimize the energy, the other coordinates will then tend to change. The quantity $(\alpha)_\beta$ tells us how coordinate α would change for minimum energy if β is constrained to unit positive displacement. These quantities have been calculated for the hexafluorides of Table 3.2-1 and are listed in Table 3.2-2.

The coordinates r_i, q_{ij}, and α_{ij} are defined by Fig. 3.2-1. The stretch–stretch interactions are relatively small and do not show any obvious

TABLE 3.2-2

INTERACTION DISPLACEMENT COORDINATES[a] FOR SOME HEXAFLUORIDES

	$(r_2)_{r_1}$	$(r_6)_{r_1}$	$(\alpha_{12})_{r_1}$	$(q_{12})_{r_1}$	$(q_{26})_{r_1}$	$(r_1)_{\alpha_{12}}$	$(\alpha_{13})_{\alpha_{12}}$	$(\alpha_{14})_{\alpha_{12}}$
RhF_6	−0.02	0.02	−0.08	0.58	0.11	−0.006	−0.06	−0.39
WF_6	−0.04	−0.09	−0.16	0.47	0.12	−0.005	−0.18	−0.21
IrF_6	−0.03	−0.02	−0.13	0.52	0.14	−0.008	−0.04	−0.42
PtF_6	−0.03	0.04	−0.12	0.53	0.16	−0.008	−0.03	−0.46
UF_6	−0.07	−0.01	0.17	0.90	−0.29	0.008	−0.06	−0.36
NpF_6	−0.06	0.02	0.08	0.78	−0.14	0.005	−0.02	−0.40
PuF_6	−0.06	0.03	0.08	0.77	−0.13	0.005	−0.02	−0.42
SF_6	−0.05	−0.09	−0.18	0.47	0.10	−0.04	−0.06	−0.35
SeF_6	−0.02	−0.05	−0.18	0.48	0.16	−0.02	−0.07	−0.34
TeF_6	−0.01	−0.04	−0.16	0.50	0.17	−0.01	−0.10	−0.33

[a] The meaning of these coordinates, as previously defined is exemplified as follows: the fact that $(r_2)_{r_1} = -0.02$ for RhF_6 means that if one RhF bond is constrained to a small positive displacement minimum energy is obtained by a contraction of the cis RhF bonds equal to 2% of the constrained displacement.

trend among these ten molecules. It does appear that the cis stretch–stretch interaction is significantly greater than the trans stretch–stretch interaction for the actinide metal fluorides. The explanation for this is not obvious.

The quantities $(q_{26})_{r_1}$ show how the F_2–F_6 distance would change for minimum energy if r_1 is stretched. If F–F repulsion were an important factor in determining the equilibrium configuration, as hypothesized for a Urey–Bradley Force Field (or the OVFF used by Kim et al. (*16*)), one would expect $(q_{26})_{r_1}$ to be positive. This is true for most of the hexafluorides but not for the actinide molecules. Therefore, at least for these latter molecules, it appears that F–F repulsion is not a major contribution to the potential energy.

Another point of interest is the quantity $(\alpha_{14})_{\alpha_{12}}$ (last column of Table 3.2-2). In first approximation, if α_{12} is increased by one unit for minimum

energy, we might expect $(\alpha_{14})_{12} = (\alpha_{26})_{12} \approx (\alpha_{46})_{\alpha_{12}} \approx 0.33$ (because of the first-order redundancy $\Delta\alpha_{12} + \Delta\alpha_{14} + \Delta\alpha_{26} + \Delta\alpha_{46} = 0$). However, for WF_6, $(\alpha_{14})_{\alpha_{12}}$ is quite low and for PtF_6 it is quite high. The reason for this is not known but may indicate some discrepancy in the force constant calculation. The Coriolis constants derived from band contours have sometimes given anomalous results (*19*) and so it is desirable to have supplementary data. A determination of more accurate potential constants, perhaps using isotopic frequency shifts, centrifugal distortion

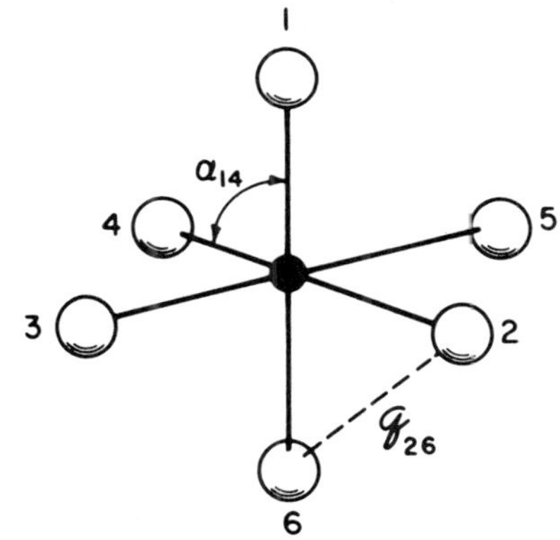

Fig. 3.2-1 Internal coordinate for octahedral XY_6.

data, etc., would no doubt be useful in pinpointing the potential constants.

As an example of the use of isotopic frequency shifts for calculation of force constants of octahedral hexafluoride, calculations have been made for several isotopic pairs of MF_6. The frequency shifts as a function of the force constants are given in Table 3.2-3.

It is apparent from this table that for the isotopic pair $^{180}WF_6$–$^{186}WF_6$ the isotope shift is too insensitive to be useful for determination of force constants. If we measured $\Delta\nu_3$ to ± 0.2 cm^{-1}, we would know F_{34} to ± 0.12 mdyn Å^{-1}. F_{33} and F_{44} would carry a corresponding uncertainty as indicated by Table 3.2-3. For $^{92}MoF_6$–$^{100}MoF_6$, the situation is much better. A knowledge of $\Delta\nu_3$ to ± 0.2 cm^{-1} gives F_{34} to ± 0.03 mdyn Å^{-1}. For the lighter molecules listed, the situation is even better. Perhaps the best is for NiF_6^{-2} for which a knowledge of $\Delta\nu_3$ to ± 0.2 cm^{-1} leads to determination of F_{34} to ± 0.013 mdyn Å^{-1}. Thus we see that studies of isotope shifts of ν_3 for the hexafluorides would be quite useful for estimation of force constants provided the mass change is great enough but the mass is not too large. It is not possible to derive a simple expression for the general case, but the sensitivity of the isotope shift on the force constants is dependent to a considerable extent on the ratio $(M^i - M^j)/M^iM^j$ where M^α is the mass of isotope α.

One other hexafluoride should be mentioned, namely, XeF_6. Electron diffraction studies (*20*, *21*) of the vapor indicate that the molecule does not have octahedral symmetry. The vibrational spectra are incomplete

TABLE 3.2-3

Relation of Force Constants[a] to Isotopic Frequency Shifts for Some Isotopic Pairs of MF_6

	F_{34}	F_{33}	F_{44}	$\lvert\Delta\nu_3\rvert$	$\lvert\Delta\nu_4\rvert$
$^{180}WF_6$	0.3	4.859	0.284	1.7	1.0
	0.0	4.674	0.276	2.2	0.8
$^{186}WF_6$	−0.3	4.378	0.315	2.7	0.7
$^{92}MoF_6$	0.3	4.699	0.267	7.8	3.3
	0.0	4.369	0.267	9.8	2.5
$^{100}MoF_6$	−0.3	3.897	0.322	11.9	1.8
$^{58}NiF_6^{-2}$	0.3	3.168	0.372	12.1	4.6
$^{64}NiF_6^{-2}$	0.0	2.531	0.430	16.8	2.1
$^{74}SeF_6$	0.3	4.757	0.658	14.7	5.2
$^{82}SeF_6$	0.0	4.206	0.723	18.4	3.1
$^{32}SF_6$	1.2[b]	5.913	1.015	13.9	5.3
$^{34}SF_6$	0.9	5.214	1.031	17.2	3.0

[a] Force constants are in mdyn Å^{-1} and frequency shifts in cm^{-1}.
[b] For SF_6 no solution was possible for F_{34} near zero.

and inconclusive concerning the structure. The reader is referred to the literature (*22*) for detailed information.

Several hexachloride, -bromide, and -iodide complexes have been studied, primarily as solids. Perhaps the most detailed calculations on these complexes are those of Hiraishi et al. (*23*) in which they calculated force constants for $PtCl_6^{-2}$, $PdCl_6^{-2}$, $SnCl_6^{-2}$, $SeCl_6^{-2}$, $PtBr_6^{-2}$, and $SnBr_6^{-2}$ as ionic solids. However, for none of these have all six fundamental frequencies been observed. For three of the complexes, five fundamentals were observed, so they were able to fit them with a modified Urey–Bradley force field having five force constants. The frequencies are given in Table 3.2-4 and the force constants in Table 3.2-5.

The force constants given in Table 3.2-5 are defined in a footnote to the table. The constant h was included by Hiraishi et al. (*23*) quite

TABLE 3.2-4

FUNDAMENTAL FREQUENCIES[a] OF SOME HEXAHALIDES

Assignment		Activity[b]	Mode[c]	K_2PtCl_6	K_2PdCl_6	K_2PtBr_6
A_{1g}	ν_1	R	ν_{MX}	344	317	207
E_g	ν_2	R	ν_{MX}	320	292	190
F_{1u}	ν_3	IR	ν_{MX}	340	340	244
	ν_4	IR	δ_{XMX}	188	175	90
F_{2g}	ν_5	R	δ_{XMX}	160	164	97
F_{2u}	ν_6	ia	δ_{XMX}	(60)[d]	(98)[d]	(68)[d]

[a] Values are in cm^{-1}.

[b] R is for Raman activity, IR for infrared, and ia for inactive.

[c] $\nu_{MX} = MX$ stretching; $\delta_{XMX} = XMX$ bending.

[d] Values for ν_6 have not been observed. Those given here are calculated from the force constants of Table 3.2-5.

arbitrarily. This corresponds to $F'_{\alpha\alpha}$ as defined in Appendix VIII. There seems to be no reason to select this interaction constant as important and neglect others; thus, the solution of Table 3.2-5 must be considered rather arbitrary.

The symmetry force constants can be expressed in terms of the Urey–Bradley force constants of Table 3.2-5. This is done in Table 3.2-6.

The Urey–Bradley force constants of Table 3.2-5 can be converted to internal valence force constants using Table 3.2-6 and Eqs. (3.2-6)–(3.2-12).

TABLE 3.2-5

MODIFIED UREY–BRADLEY FORCE CONSTANTS FOR SOME HEXAHALIDE COMPLEXES

	$PtCl_6^{-2}$	$PdCl_6^{-2}$	$PtBr_6^{-2}$
K[a]	1.86	1.51	1.54
F	0.10	0.096	0.096
H	0.077	0.087	0.057
h	0.047	0.020	−0.005
k	0.21	0.207	0.096

[a] K, H, and F represent MX-stretching, XMX-bending, and X—X force constants, respectively. The constant k is for interaction of two opposite MX bonds. The constant h is for interaction of two angles, one bond of each being shared while the other bonds are at 90° to each other. All values are in mdyn $Å^{-1}$.

TABLE 3.2-6

SYMMETRY FORCE CONSTANTS OF OCTAHEDRAL MX_6 AS FUNCTIONS OF MODIFIED UREY–BRADLEY FORCE CONSTANTS

$$F_{11} = K + k + 4F$$
$$F_{22} = K + k + 0.7F$$
$$F_{33} = K - k + 1.8F$$
$$F_{34} = 0.9F$$
$$F_{44} = H + 0.55F + 2h$$
$$F_{55} = H + 0.55F$$
$$F_{66} = H + 0.55F - 2h$$

From the results the interaction displacement coordinates can be calculated using Eqs. (VIII-3)–(VIII-8) from Appendix VIII. The results are given in Table 3.2-7.

The results of Table 3.2-7 indicate some anomalies. For example, the sign change of $(\alpha_{13})_{\alpha_{12}}$ for $PdCl_6^{-2}$ gives doubt to the accuracy of the force field. However, in general, the values are reasonable. It is interesting that the term $(\alpha_{12})_{r_1}$ is of greater magnitude for $PtBr_6^{-2}$ than for $PtCl_6^{-2}$. This is in accord with importance of halide–halide repulsion as included in the Urey–Bradley Force Field because the Br–Br distance is about 0.4 Å less than the van der Waals diameter while the Cl–Cl distance is only 0.3 Å less than the van der Waals diameter.

The results discussed in this section give some support to the modified Urey–Bradley Force Field. However, the evidence is not conclusive and further data, including isotopic frequency shifts and Coriolis constant evaluation, where possible, is necessary for verification.

TABLE 3.2-7

INTERACTION DISPLACEMENT COORDINATES FOR SOME HEXAHALIDES

Coordinate[a]	$PtCl_6^{-2}$	$PdCl_6^{-2}$	$PtBr_6^{-2}$
$(r_2)_{r_1}$	−0.05	−0.02	−0.03
$(r_6)_{r_1}$	−0.11	−0.13	−0.08
$(\alpha_{12})_{r_1}$	−0.05	−0.06	−0.09 rad Å^{-1}
$(r_1)_{\alpha_{12}}$	−0.04	−0.03	−0.05 Å rad^{-1}
$(\alpha_{13})_{\alpha_{12}}$	+0.07	−0.10	+0.04
$(\alpha_{14})_{\alpha_{12}}$	−0.36	−0.31	−0.32

[a] See footnote to Table 3.2-2 and Fig. 3.2-1 for a description of these coordinates.

3.3. Tetrahalide Complexes of D_{4h} Symmetry

There are relatively few known complex ions or molecules which fall in this class. Some of these are $PtCl_4^{-2}$, $PdCl_4^{-2}$, $AuCl_4^{-1}$, and the corresponding bromide complexes. The molecules XeF_4 and perhaps ClF_4^- are the only square planar fluoride complexes known so far. The number and type of vibrations in the various representations of D_{4h} can readily be obtained

TABLE 3.3-1

Symmetry Force Constants of MX_4 of D_{4h} Symmetry

A_{1g} (R)[a]	$F_{11} = F_r + 2F_{rr}^c + F_{rr}^t$ [b]
B_{1g} (R)	$F_{22} = F_r - 2F_{rr}^c + F_{rr}^t$
B_{2g} (R)	$F_{33} = F_\alpha - 2F'_{\alpha\alpha} + F''_{\alpha\alpha}$
A_{2u} (I)	$F_{44} = F_\epsilon + F'_{\epsilon\epsilon}$
B_{2u}	$F_{55} = F_\epsilon - F'_{\epsilon\epsilon}$
E_u (I)	$F_{66} = F_r - F_{rr}^t$
	$F_{77} = F_\alpha - F''_{\alpha\alpha}$
	$F_{67} = 2^{1/2}(F_{r\alpha} - F'_{r\alpha})$

[a] R means Raman active and I means infrared active.

[b] The coordinates are defined as, $r = MX$ bond, α = in plane XMX angle, ϵ = out of plane XMX angle. The interaction constants are the same as those defined in Appendix II.

from the results on $M(CN)_4^{-x}$ in Appendix II by disregarding the N atom. Thus, for MX_4, from Table II-2 (Appendix II), we find that the various vibrations are distributed as

$$1A_{1g}\,(r_{MX}) + 1B_{1g}\,(r_{MX}) + 1B_{2g}\,(\alpha_{XMX}) + 1A_{2u}\,(\epsilon_{XMX}) + 1B_{2u}\,(\epsilon_{XMX}) + 2E_u\,(r_{MC} + \alpha_{CMC})$$

There is also a redundant coordinate, $A_{1g}\,(\alpha_{XMX})$, which does not contribute to the vibrational spectrum. The effect of the redundancy on the meaning of the force constants is discussed by Jones (*24*).

The symmetry force constant matrix for a General Quadratic Valence Force Field can be extracted from Table II-6 by ignoring the N atom. The results are given in Table 3.3-1.

There are seven observable frequencies for MX_4 but eight force constants. The symmetry force constants F_{11} through F_{55} can be calculated directly from the observed frequencies ν_1 through ν_5. Some of the internal valence force constants can then be calculated as shown in the following equations.

$$F_{rr}^c = (F_{11} - F_{22})/4 \tag{3.3-1}$$
$$F_r + F_{rr}^t = (F_{11} + F_{22})/2 \tag{3.3-2}$$
$$F_\epsilon = (F_{44} + F_{55})/2 \tag{3.3-3}$$
$$F_{\epsilon\epsilon}' = (F_{44} - F_{55})/2 \tag{3.3-4}$$

However, for the E_u species there are only two frequencies to calculate three force constants. Either more information is needed or a more

TABLE 3.3-2

FUNDAMENTAL VIBRATIONAL FREQUENCIES OF MX_4

	$AuCl_4^-$	$AuBr_4^-$	$PtCl_4^{-2}$
A_{1g} ν_1[a]	347	212	335
B_{1g} ν_2	324	196	304
B_{2g} ν_3	171	102	164
A_{2u} ν_4	144	—	167
E_u ν_6	356	252	316
ν_7	173	100	185

[a] These are numbered in conformity with Table 3.3-1. In (*25*) a different numbering system is used.

restrictive potential function must be used to complete the force constant calculation.

Sabatini et al. (*25*) report the five in-plane fundamentals for $AuCl_4^-$, $AuBr_4^-$, and $PtCl_4^{-2}$. One out-of-plane frequency is given for $AuCl_4^-$ and $PtCl_4^-$. These results are given in Table 3.3-2. Sabatini et al. (*25*) achieved a solution by making the arbitrary assumption that $F_{67} = 0$. They also showed that the Urey–Bradley Force Field did not give a satisfactory fit.

Since that time, Hiraishi and Shimanouchi (*26*) have treated K_2PtCl_4, including the interaction of the K^+ ions and the $PtCl_4^{-2}$ ions in the crystal lattice. These interesting results are shown in Table 3.3-3.

The force constants of Set I yield the frequencies of Set I which agree reasonably well with the observations. If we use the same force constants except that $K(KX)$ is neglected, we arrive at the results of Set II. Note that the A_{2u} and E_u bending frequencies are severely affected. This shows that when one neglects the lattice frequencies and force constants in a normal coordinate treatment using solid state frequencies, the results may be subject to considerable error.

A modified Urey–Bradley force field is used in this treatment; however, the modification (inclusion of P, P', and h') is so extensive as to make the significance of the Urey–Bradley constant questionable.

TABLE 3.3-3

FORCE CONSTANTS AND FREQUENCIES OF SOLID K_2PtCl_4

	Force Constants (mdyn Å^{-1})			Frequencies (cm^{-1})		
	Set I	Set II		Obs.	Set I	Set II
$K(MX)$[a]	1.817	1.817	A_{1g}	335	335	335
P	0.270	0.270	B_{1g}	304	304	304
P'	0.043	0.043	B_{2g}	164	164	164
$H(XMX)$	0.105	0.105	A_{2u}	170	175	123
h'	0.013	0.013		103[c]	100	0
$F(XX)$	0.11[b]	0.11[b]	E_u	325	325	325
$P(MX)$	0.046	0.046		190	192	142
$K(KX)$	0.115	0		116[c]	110	0
				90[c]	87	0

[a] The definition of force constants is: $K(MX)$ = MX stretching constant; P = interaction of opposite MX bonds; P' = interaction of adjacent MX bonds; $H(XMX)$ = in-plane angle bending constant; h' = interaction of adjacent in-plane angles; $F(XX)$ = halide–halide force constant; $P(MX)$ = out-of-plane bending constant; and $K(KX)$ = force constant for $K^+ - Cl^-$ interaction.

[b] Fixed on basis of argon gas repulsive potential.

[c] These are lattice frequencies.

The infrared and Raman spectra of XeF_4 have been observed (*27*) and are in agreement with a planar D_{4h} structure as found in crystal structure studies (*28*). The fundamental frequencies of the gas are now assigned (*28*) as $\nu_1 = 550$ cm^{-1}, $\nu_2 = 519$, $\nu_3 = 235$, $\nu_4 = 291$, $\nu_5 \approx 246$, $\nu_6 = 586$, and $\nu_7 \approx 246$. The numbering of frequencies corresponds to Table 3.3-1. It is different from that in (*27*) and (*28*), in which the representation assignment differs from the usual character tables. The frequencies ν_5 and ν_7 were assigned (*28*) by a very indirect method (through thermodynamic functions) and, thus, cannot be considered reliable. Therefore, it is not appropriate to attempt a complete force constant calculation at this stage. However, from ν_1 and ν_2 the constants F_{rr}^c and $F_r + F_{rr}^t$ are calculated as 0.093 and 3.20 mdyn Å^{-1} using Eqs. (3.3-1) and (3.3-2). Thus, the Xe-F force constant is about 3 mdyn Å^{-1}, considerably weaker than the Te-F bond in TeF_6 (5.1 mdyn Å^{-1}). This is in agreement with Rundle's theory (*29*) of bonding for XeF_4, in which he proposes a 3 center–4 electron bond which should be somewhat weaker than normal dsp^2 or d^2sp^3 bonds.

3.4. Tetrahedral Halide Complexes

The group of metals which forms tetrahedral complexes is much larger than the group which forms square planar complexes. The vibrational

TABLE 3.4-1

Fundamental Frequencies and Selection Rules for MX_4

Square Planar (D_{4h})			Tetrahedral (T_d)		
A_{1g}	ν_1	(R)[a]	A_1	ν_1	(R)
B_{1g}	ν_2	(R)	E	ν_2	(R)
B_{2g}	ν_3	(R)	F_2	ν_3	(R, I)
				ν_4	(R, I)
A_{2u}	ν_4	(I)			
B_{1u}	ν_5	(in)			
E_u	ν_6	(I)			
	ν_7	(I)			

[a] The letters in parentheses denote the selection rules. Thus, R is for Raman active; I is for infrared active; and *in* means inactive in both infrared absorption and Raman.

spectrum of the tetrahedral complexes has not been studied as extensively. It should be possible to distinguish between the two from their vibrational spectra as the number of distinct fundamental frequencies and their selection rules are different. This is shown in Table 3.4-1.

As shown in the previous section, for $AuCl_4^-$, $AuBr_4^-$, and $PtCl_4^{-2}$, six frequencies have been observed—three in the Raman and three in the infrared. The infrared peaks do not coincide with Raman peaks. Any near coincidences observed can be considered accidental. Thus the number and selection rules for the observed frequencies are in agreement with D_{4h} symmetry and not T_d symmetry.

Sabatini and Sacconi (*30*) have studied the infrared spectra from 400 to 70 cm^{-1} for a large number of compounds of the type L_2MX_4, in which L is Me_4N^+, Et_4N^+, or Pr_4N^+ and X is Cl, Br, or I. The metals studied were Mn, Fe, Co, Cu, Zn, and Ni, all in the divalent state. Unfortunately,

Raman studies of these compounds have not been made except for $ZnCl_4^{-2}$ and $ZnBr_4^{-2}$ in aqueous solution (*31*). The results of (*30*) and (*31*) are given in Table 3.4-2.

For $ZnCl_4^{-2}$ and $ZnBr_4^{-2}$, the Raman studies and infrared studies were of different phase (solution and solid, respectively). Therefore, we cannot expect exact coincidences for the F_2 vibrations (ν_3 and ν_4) in the Raman and infrared spectra. They are close enough to presume that they arise from the same vibration, thus suggesting the tetrahedral structure. Reliable force constants cannot be calculated from this data as there are

TABLE 3.4-2

VIBRATIONAL FREQUENCIES OF $ZnCl_4^{-2}$ AND $ZnBr_4^{-2}$

	$ZnCl_4^{-2}$		$ZnBr_4^{-2}$	
	Raman[a]	IR[b]	Raman[a]	IR[b]
$\nu_1(A_1)$	282		172	
$\nu_2(E)$	82		61	
$\nu_3(F_2)$		276	210	207
$\nu_4(F_2)$	116	133	82	92

[a] Aqueous solution (*31*).
[b] Solid state, L_2MX_4 (*30*).

five force constants but only four frequencies; however, it is clear that the MX stretching frequencies are considerably lower than those of square planar complexes involving the same halide (see Sect. 3.3). Thus the tetrahedral M—X bonds are observed to be considerably weaker than square planar M—X bonds.

There are, of course, a number of tetrahedral molecules, in which the center atom is not a metal, which have been studied more thoroughly. An extensive list of these, with the observed frequencies and references, is given by Nakamoto (*32*). As there are five force constants and only four frequencies, unique general quadratic force constant solutions have not been obtained except in a few cases where other data, such as isotopic shifts and Coriolis coupling constants, are available. In a very careful study of methane, Duncan and Mills (*33*) state that "frequency data alone is not sufficient to determine the General Force Field" for the F_2 species of methane. They show that Coriolis zeta constants with frequency data do indeed determine the force field. However, their results are in close agreement with those determined (*34*) from frequency data alone on CH_4, CD_4, and CT_4. This agreement arises only because harmonic frequencies (observed frequencies corrected for anharmonicity) were used; if the uncorrected observed frequencies are used a considerably different result is obtained; whereas, using the Coriolis zeta values with the uncorrected frequencies gives results closer to the true, harmonic results. The conclusion is that if the anharmonic corrections are large and unknown, the Coriolis zeta values are much more helpful than the isotopic frequency shifts in determining the true force field. However, if the anharmonicity corrections are known, or not very large, the isotopic frequency shifts are very useful in determining the force field. Chalmers and McKean (*35*)

TABLE 3.4-3

HARMONIC FORCE CONSTANTS OF METHANE

F_r	5.50 mdyn Å^{-1}
F_{rr}	0.11 mdyn Å^{-1}
$h_\alpha - h'_{\alpha\alpha}$ [a]	0.56 mdyn Å rad^{-2}
$h_\alpha - F''_{\alpha\alpha}$ [a]	0.55 mdyn Å rad^{-2}
$h_{r\alpha}$ [a]	0.16 mdyn rad^{-1}

[a] These force constants are defined in Sect. 1.10.

have discussed the effect of anharmonicity in detail and have shown that for small isotope shifts the uncertainty about anharmonicity is much less important than for large shifts (such as AH → AD). The harmonic force constants for methane are given in Table 3.4-3.

From Eqs. (1.10-13) and (1.10-14), we can calculate the interaction displacement coordinates: $(r_2)_{r_1} = -0.023$ and $(\alpha_{12})_{r_1} = -0.15$ rad Å^{-1}. This means that stretching r_1 a small amount leads for minimum energy to a contraction of r_2 equal to 2.3% of that amount and a decrease in the adjacent angles by 0.15 rad Å^{-1} of stretch of r_1. Actually this is in line with the ideas of Mills (*36*) concerning a Hybrid Orbital Force Field. As applied to methane, this field implies that in S—P bonds greater S character leads to shorter bonds and larger angles. Thus pairs of SP bonds are linear, SP^2 bonds are at 120°, SP^3 bonds at 109°28′, and P^2 bonds are at 90°. Stretching a CH bond should give it greater P character. The other three CH bonds will then have more S and therefore tend to become shorter. Also the angles between the other three CH bonds will become greater because they have more S character. This leads qualitatively to the interaction displacement coordinates calculated above.

3.5. Pentacoordinated Species

Coordination of a central atom with five similar ligands results in two arrangements: the trigonal bypyramid (Fig. 3.5-1a) and the tetragonal pyramid (Fig. 3.5-1b). If, after the five *M—F* bonds are formed, there are no extra valence shell electrons, the trigonal pyramid appears to be the most stable structure. The theory of the bonding in pentacoordinated species is still not settled; for example, Gillespie (*37*) and Bartell (*38*) discuss the bonding in PF_5, $(CH_3)PF_4$, and $(CH_3)_2PF_3$ from quite different

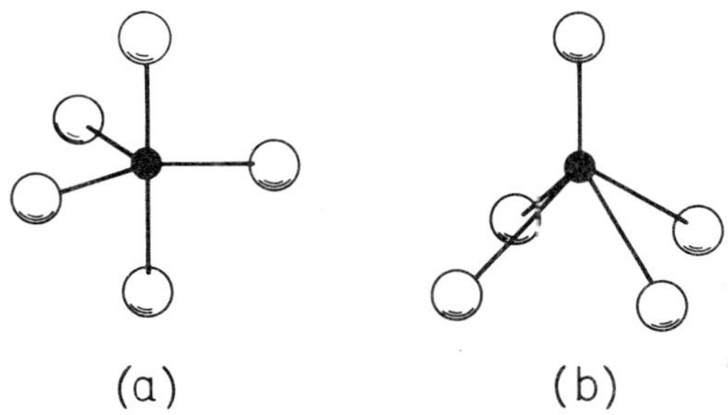

Fig. 3.5-1 Schematic diagrams of (a) trigonal bypyramid and (b) tetragonal pyramid.

approaches (electron pair repulsion and molecular orbitals, respectively) to explain the observed results. Some of the molecules found to have D_{3h} symmetry are PF_5, SbF_5, PCl_5, $SbCl_5$, $NbCl_5$, $TaCl_5$, and VF_5.

On the other hand, molecules like ClF_5, BrF_5, and IF_5 have the tetragonal pyramid (C_{4v}) structure because they have a lone pair of electrons in the valence shell which must occupy one of the positions of a hexacoordinated structure.

The D_{3h} MX_5 molecule has eight fundamental vibrations as indicated in Table 3.5-1.

The results for $NbCl_5$ and $TaCl_5$ are not given here as they were considered quite tentative.

In the general quadratic force field, there are thirteen force constants to calculate from eight frequencies. However, force constants have been calculated for the first four molecules in Table 3.5-1 using a modified Urey–Bradley force field (*39*). The results are given in Table 3.5-2. Similar constants were calculated for VF_5 but the frequency fit was not very good so they are not reproduced here.

TABLE 3.5-1

FREQUENCIES[a] FOR MX_5 OF D_{3h} SYMMETRY

		PF_5[b]	PCl_5[b,c]	SbF_5[b]	$SbCl_5$[b]	VF_5[d]
A_1' (R)[e]	ν_1	817	394	718	356	719
	ν_2	640	264	672	307	608
A_2' (I)	ν_3	945	465	724	384	784
	ν_4	576	299	212	154	331
E' (R, I)	ν_5	~1028	592	758	398	810
	ν_6	534	273	265	172	282
	ν_7	301	100	93	74	(~200)
E'' (R)	ν_8	514	282	224	165	350

[a] Values are in cm^{-1}.
[b] Taken from (39). References to other work can be found in (*39*).
[c] The values for PCl_5 from (*39*) are somewhat different from those given by Carlson (*40*). However, they are more recent and apparently preferable.
[d] From (*41*).
[e] R stands for Raman active and I for infrared active.

TABLE 3.5-2

MODIFIED UREY–BRADLEY FORCE CONSTANTS[a] FOR $MX_5(D_{3h})$

	PF_5	PCl_5	SbF_5	$SbCl_5$
K_r[a]	5.50	2.10	5.30	2.20
K_d	3.85	1.00	4.60	1.83
H_α	0.37	0.07	0.095	0.06
H_β	0.48	0.15	0.12	0.085
F_{rr}	0.15	0.10	0.07	0.08
F_{dr}	0.80	0.55	0.10	0.13
$H_{\alpha\beta}$	0.09	0.07	0.07	0.043

[a] The force constants are defined as: K_r (stretch of equatorial bond); K_d (stretch of axial bond); H_α (bend of equatorial YXY angle); H_β (bend of angle between equatorial and axial bond); F_{rr} (constant for equatorial X–X stretch); F_{dr} (stretch of equatorial X from axial X); $H_{\alpha\beta}$ (interaction of equatorial angle with axial angle). All values are in mdyn Å^{-1}.

The force constant solution is not unique though the fact that eight frequencies are fit fairly well with seven force constants suggests that the Urey–Bradley field may be appropriate for these molecules. There are some anomalies, however. Recently Hansen and Bartell (*42*) found that for PF_5 the axial PF distance (1.577 Å) is longer than the equatorial PF distance (1.534 Å). This agrees with the larger value for K_r than for K_d given in Table 3.5-2. However, the difference is somewhat greater than one might expect for a 2.8% change in distance. Furthermore, the ratio of equatorial to axial bond force constants for PCl_5 is greater than two, somewhat greater than one would expect from the ratio of distance (*43*) in PCl_5 vapor (2.19/2.04). Using Badger's rule (*44*), one would expect this ratio of force constants to be about 1.7 rather than 2.1. It would be interesting to have estimates of Coriolis zeta values to verify these force constants.

The interhalogens ClF_5, BrF_5, and IF_5 are thought to form tetragonal pyramids. Infrared and Raman studies have been reported (*45*, *46*) and force constants calculated (*46*). For the tetragonal pyramid of C_{4v} symmetry, there are nine fundamental frequencies ($3A_1$, $2B_1$, $1B_2$, and $3E$). They are all Raman active but only A_1 and E are infrared active. There are thirteen force constants to calculate from nine frequencies. Begun and Fletcher (*46*) calculated force constants; however, their necessary assumptions were quite arbitrary and therefore little meaning can be attached to the results except in a qualitative fashion. It is remarkable, however, that the *M-F* stretching force constants decrease in the order $F_{IF} > F_{BrF} > F_{ClF}$. Ordinarily we would expect the lighter atom to form the stronger bond because it can be shorter. It would be interesting to determine these force constants more uniquely and to compare them with theories of bonding concerning such molecules.

3.6. Pyramidal Trifluorides of N, P, and As

Mirri (*47*) has calculated force constants for NF_3, PF_3, and AsF_3 using centrifugal distortion constants (*48–50*), Coriolis coupling constants (*51*), and vibrational frequencies (*52*), uncorrected for anharmonicity. The results are given in Table 3.6-1.

The observed values of ζ_4 were taken from band contour measurements (*51*). They are (*51*) −0.965, −0.64, and −0.435 for NF_3, PF_3, and AsF_3, respectively. It is interesting to note that ζ_4 for NF_3 was determined from high resolution studies (*53*) to be −0.90. Thus the band contour method is in error by 7% in this case. The value settled on by Mirri

(−0.93) is right in between the approximate and more exact values. It should be pointed out that the use of anharmonic frequencies may lead to some error in the results of Table 3.6-1 and this is not included in the uncertainties given.

The symmetry force constants have been converted to internal coordinate force constants by the author and are given in Table 3.6-2. The F-F distance is given also as it is pertinent to the interactions.

TABLE 3.6-1

FORCE CONSTANTS[a], CORIOLIS COUPLING CONSTANTS AND CENTRIFUGAL DISTORTION[a] CONSTANTS FOR NF_3, PF_3, AND AsF_3

	NF_3	PF_3	AsF_3
F_{11}	6.29 ± 0.14[b]	6.18 ± 0.10	4.80 ± 0.06
F_{22}	2.34 ± 0.04	2.01 ± 0.02	1.66 ± 0.04
F_{12}	0.88 ± 0.04	0.49 ± 0.09	-0.48 ± 0.10
F_{33}	3.22 ± 0.07	4.96	4.34 ± 0.01
F_{44}	1.75 ± 0.03	1.20	0.918 ± 0.001
F_{34}	-0.36 ± 0.04	−0.32	-0.23 ± 0.01
ζ_4	-0.93 ± 0.01	-0.60_6	$-0.45_5 \pm 0.00_5$
D_J	14.53 ± 0.05	78.45 ± 0.5	4.62 ± 0.02
D_{JK}	-22.80 ± 0.10	−11.88	-6.17 ± 0.05

[a] Units of force constants are mdyn $Å^{-1}$ for F_{11} and F_{33}, mdyn rad^{-1} for F_{12} and F_{34}, mdyn Å rad^{-2} for F_{22} and F_{44}. For D_J and D_{JK} units are Kcps. Values are from *(47)*.

[b] Uncertainties in the force constants were obtained *(47)* by requiring that they be consistent with the *CD* constants within their experimental error and with the ζ_4 parameter within the accuracy of 5%. They are not standard deviations.

The close proximity of the fluorine atom in NF_3 may explain the fact that the N*F* stretching constant is weaker than that for PF stretching. The F–F repulsion will increase the N–F bond length and thus weaken it. The mutual interaction of *XF* bonds may also arise from F–F repulsions; the dramatic decrease from NF_3 to PF_3 to AsF_3 would be difficult to explain otherwise. The relatively large negative value for $F_{XF,\alpha'}$ for AsF_3 is inconsistent with the others. It implies that stretching an AsF bond would tend to cause contraction of the opposite angle. This is possible because the F–F repulsion is probably very small for this molecule; however, it may also indicate some discrepancies in the results.

The fact that F–F repulsion appears to be important for NF_3 and to a lesser extent PF_3 suggests that a Urey–Bradley force field might be

TABLE 3.6-2

VALENCE FORCE CONSTANTS OF NF_3, PF_3, and AsF_5

	NF_3	PF_3	AsF_3
F_{XF}	4.24	5.37	4.49 mdyn Å^{-1}
F_{α}	1.95	1.47	1.17 mdyn Å rad^{-2}
$F_{XF,XF'}$	1.02	0.41	0.15 mdyn Å^{-1}
$F_{\alpha\alpha'}$	0.20	0.27	0.25 mdyn Å·rad^{-2}
$F_{XF,\alpha}$	0.41	0.27	−0.08 mdyn rad^{-1}
$F_{XF,\alpha'}$	0.06	−0.05	−0.31 mdyn rad^{-1}
R_{F-F}	2.1	2.4	2.7 Å

TABLE 3.6-3

RELATION OF UREY–BRADLEY CONSTANTS[a] TO QUADRATIC SYMMETRY FORCE CONSTANTS FOR PYRAMIDAL XY_3 MOLECULES

	Symmetry force constants	Urey–Bradley constants
Species A_1	F_{11}	$K + 4s^2F$
	F_{22}	$H - s^2F' + t^2F$
	F_{12}	$2st(F + F')$
Species E	F_{33}	$K + 3t^2F' + s^2F$
	F_{44}	$H - s^2F' + t^2F$
	F_{34}	$-st(F + F')$

[a] K is the X–Y force constant, H for YXY bending, F for Y–Y stretching, and F' for Y–Y repulsive constant. $s^2 = \frac{1}{2}(1 - \cos XYX)$; $c^2 = \frac{1}{2}(1 + \cos YXY)$.

TABLE 3.6-4

VALENCE FORCE CONSTANTS[a] OF NH_3

	MUBFF	GQVF
F_r	6.969	6.920
F_{α}/r^2	0.666	0.660
F_{rr}	0.006	0.049
$F_{\alpha\alpha'}/r^2$	0	−0.020
$F_{r\alpha}/r$	0.405	0.496
$F_{r\alpha'}/r$	0.409	0.301

[a] Values are in mdyn Å^{-1}.

appropriate for these molecules. The relations have been given by Meisingeth (*53*). They are reproduced in Table 3.6-3.

It is seen in Table 3.6-3 that if the Urey–Bradley force field applies, $F_{22} = F_{44}$ and $F_{12} = -2F_{34}$. From Table 3.6-1, we see that the latter equality is approximately held for NF_3 and PF_3 but in no sense by AsF_3. For all three molecules F_{22} is significantly greater than F_{44}, showing that the Urey–Bradley force field requires modification to be applicable.

Pariseau et al. (*54*) have shown that for NH_3, PH_3, SbH_3, and AsH_3, a modification of the Urey–Bradley force field, based on consideration of the lone pair of electrons, is a useful force field. Table 3.6-4 gives a comparison of the Valence Force Constants as determined from the Modified Urey–Bradley force field (MUBFF) with the general quadratic valence force constants (GQVF) (*54*) for NH_3. The results for SbH_3 and AsH_3 are equally good while for PH_3 the agreement is poorer, presumably because the frequency data is less reliable (*54*). Perhaps a similar treatment would prove satisfactory for the pyramidal trifluorides.

REFERENCES

1. L. P. Lindeman and M. K. Wilson, *J. Chem. Phys.*, **24,** 242 (1956) and earlier papers.
2. C. W. F. T. Pistorius, *J. Chem. Phys.* **29,** 1174 (1958).
3. J. A. Ladd, W. J. Orville-Thomas, and B. C. Cox, *Spectrochim. Acta*, **19,** 1911 (1963).
4. I. W. Levin and S. Abramowitz, *J. Chem. Phys.*, **43,** 4213 (1965).
5. D. C. McKean, *Spectrochim. Acta*, **22,** 269 (1966).
6. S. Konaka, Y. Murata, K. Kuchitsu, and Y. Morino, *Bull. Chem. Soc. Japan*, **39,** 1134 (1966).
7. T. Shimanouchi, I. Nakagawa, J. Hiraishi, and M. Ishii, *J. Mol. Spectry.*, **19,** 78 (1966).
8. N. Dreska, K. N. Rao, and L. H. Jones, *J. Mol. Spectry.*, **18,** 404 (1965).
9. J. L. Duncan, *J. Mol. Spectry.*, **13,** 338 (1964); *ibid.*, **22,** 247 (1967).
10. T. Wentink, Jr., and V. H. Tiensuu, *J. Chem. Phys.*, **28,** 826 (1958).
11. D. M. Gage and E. F. Barker, *J. Chem. Phys.*, **7,** 455 (1939).
12. J. Vanderryn, *J. Chem. Phys.*, **30,** 331 (1959).
13. L. H. Jones, unpublished results.
14. E. Meisingseth, *Acta Chem. Scand.*, **16,** 1601 (1962).
15. H. H. Claassen, *J. Chem. Phys.*, **30,** 968 (1959).
16. H. Kim, P. A. Souder, and H. H. Claassen, *J. Mol. Spectry.*, **26,** 46 (1968); C. W. F. T. Pistorius, *J. Chem. Phys.*, **29,** 1328 (1959).
17. S. Abramowitz and I. Levin, *J. Chem. Phys.*, **44,** 3353 (1966).
18. L. H. Jones, *J. Chem. Phys.*, **23,** 2105 (1955).
19. R. S. McDowell, *Inorg. Chem.*, **6,** 1759 (1967).
20. L. S. Bartell, R. M. Gavin, Jr., H. B. Thompson, and C. L. Chernick, *J. Chem. Phys.*, **43,** 2547 (1965).

21. K. Hedberg, S. H. Peterson, R. R. Ryan, and B. Weinstock, *J. Chem. Phys.*, **44**, 1726 (1966).
22. B. Weinstock, E. E. Weaver, and C. P. Knop, *Inorg. Chem.*, **5**, 2189 (1966); and *Noble Gas Compounds* (H. H. Hyman, ed.), Univ. Chicago Press, Chicago, Illinois, 1963.
23. J. Hiraishi, I. Nakagawa, and T. Schimanouchi, *Spectrochim. Acta*, **20**, 819 (1964).
24. L. H. Jones, *Coord. Chem. Rev.*, **1**, 351 (1966).
25. A. Sabatini, L. Sacconi, and V. Schettino, *Inorg. Chem.*, **3**, 1775 (1964).
26. J. Hiraishi and T. Shimanouchi, *Spectrochim. Acta*, **22**, 1483 (1966).
27. H. H. Claassen, C. L. Chernick, and J. G. Malm, *J. Am. Chem. Soc.*, **85**, 1927 (1963).
28. B. Weinstock, E. E. Weaver, and C. P. Knop, *Inorg. Chem.*, **5**, 2189 (1966).
29. R. E. Rundle, *J. Am. Chem. Soc.*, **85**, 112 (1963).
30. A. Sabatini and L. Sacconi, *J. Am. Chem. Soc.* **86**, 17 (1964).
31. M. L. Delwaulle, *Compt. Rend.*, **238**, 2522 (1954).
32. K. Nakamoto, *Infrared Spectra of Inorganic and Coordination Compounds*, Wiley, New York, 1963, p. 106.
33. J. L. Duncan and I. M. Mills, *Spectrochim. Acta*, **20**, 523 (1964).
34. L. H. Jones and R. S. McDowell, *J. Mol. Spectry.*, **3**, 632 (1959).
35. A. A. Chalmers and D. C. McKean, *Spectrochim. Acta*, **22**, 251 (1966).
36. I. M. Mills, *Spectrochim. Acta*, **19**, 1585 (1963).
37. R. J. Gillespie, *Inorg. Chem.*, **5**, 1634 (1966).
38. L. S. Bartell, *Inorg. Chem.*, **5**, 1635 (1966).
39. R. A. Condrate and K. Nakamoto, *Bull. Chem. Soc. Japan*, **39**, 1108 (1966).
40. G. L. Carlson, *Spectrochim. Acta*, **19**, 1291 (1963).
41. H. H. Claassen and H. Selig, *J. Chem. Phys.*, **44**, 4039 (1966).
42. K. W. Hansen and L. S. Bartell, *Inorg. Chem.*, **4**, 1775 (1965).
43. M. Rouault, *Ann. Physique* (N.Y.), **14**, 78 (1940).
44. R. M. Badger, *J. Chem. Phys.*, **2**, 128 (1934); **3**, 710 (1935).
45. R. S. McDowell and L. B. Asprey, *J. Chem. Phys.* **37**, 165 (1962).
46. G. M. Begun and W. H. Fletcher, *J. Chem. Phys.*, **42**, 2236 (1965).
47. A. M. Mirri, *J. Chem. Phys.*, **47**, 2823 (1967).
48. A. M. Mirri, F. Scappini, and P. G. Favero, *Spectrochim. Acta*, **21**, 965 (1965).
49. A. M. Mirri and G. Cazzoli, *J. Chem. Phys.*, **47**, 1[illegible]97 (1967).
50. A. M. Mirri and D. Darniani, *Atti. Accad. Nazl. Lincei., Rend., Classe Sci. Fis. Mat. Nat.*, **40**, 635 (1966).
51. L. C. Hoskins, *J. Chem. Phys.*, **45**, 4594 (1966).
52. I. W. Levin and S. Abramowitz, *J. Chem. Phys.*, **44**, 2562 (1966).
53. E. Meisingeth, *Acta Chem. Scand.*, **17**, 509 (1963).
54. M. Pariseau, E. Wu, and J. Overend, *J. Chem. Phys.*, **39**, 217 (1963).

Chapter 4

Metal Cyanide Complexes

4.1. Free and Bound Cyanide Ion

The cyanide ion forms a large number of interesting complexes with various metals. Cyanide forms strongly covalent bonds with most metals leading generally to low-spin complexes. For a long time, it was not known whether it was bonded through the carbon or the nitrogen atom, though it was generally presumed to be the former. Neutron diffraction studies of potassium cobalticyanide by Curry and Runciman (*1*) indicate that the carbon atoms, and not the nitrogen atoms, are bonded to the cobalt. A similar study of potassium zinc tetracyanide by Sequeira and Chidambaram (*2*) concludes that the cyanides are bound to the zinc through the carbon atoms. Infrared studies (*3*) of isotopic species of $Au(CN)_2^-$ point to the conclusion that the C atom is bonded to the gold. All available evidence points to bonding of cyanide to metal through the carbon atom so this will be assumed for all metal–cyanide complexes, although there are some interesting bridge complexes for which the nitrogen is weakly bonded to a metal.

Vibrational spectra are of value for studying the effect of complexing on the CN stretching force constant as well as for studying the metal–carbon bond and the various interactions. Before discussing the complexes, it is worthwhile to consider the free cyanide ion.

The CN^- ion is isoelectronic with N_2, CO, and NO^+. Their electronic levels can be qualitatively described by molecular orbital theory (*4*). Since N_2 is homonuclear its ground state is described as

$$N_2\text{:}\quad (1s\sigma^b)^2(1s\sigma^*)^2(2s\sigma^b)^2(2s\sigma^*)^2(2p\sigma^b)^2(2p\pi^b)^4$$

The superscript *b* stands for a bonding orbital while the asterisk denotes an antibonding orbital. Since antibonding orbitals essentially cancel out bonding orbitals, nitrogen has a net of three bonds; thus, we say a strong triple bond is formed. However, the situation for heteronuclear molecules is certainly more complex. The ground state of CN^- may be described (*4*) as CN^-: $(1s\sigma^b)^2(1s\sigma^*)^2(2\sigma_N)^2(2\sigma^b)^2(2\pi^b)^4(2\sigma_C)^2$. The $2\sigma_N$

Fig. 4.1-1 Nonbonded (a) and bonded (b) structures of CN^-.

orbital is a nitrogen $2s$-$2p$ hybrid orbital and its two electrons form a nonbonding lone pair on the nitrogen atom. The $2\sigma^b$ and $2\pi^b$ orbitals form a sigma bond and two π bonds leading to a triple bond. However, this bond is weaker than the N≡N triple bond because both the σ and π bonds are unsymmetrical and are polarized toward the more electronegative nitrogen atom. This follows the argument of Purcell (*5*, *6*) based on molecular orbital calculations. The $2\sigma_C$ orbital is a $2s$-$2p$ hybrid localized primarily on the carbon atom. This is of relatively high energy and thus its electrons are readily donated in sigma bond formation. A schematic representation is shown in Fig. 4.1-1.

Figure 4.1-1(a) represents the free CN^- ion for which the σ and π bonds are weakened by polarization toward the nitrogen atom. When the CN group bonds to another atom, through the lone pair $2\sigma_C$ electrons, the negative charge on the cyanide group is lost or decreased, and the C≡N bonds become more symmetrical and thus stronger. Thus, the free cyanide ion has a force constant of 16.9 mdyn Å^{-1} (see Appendix V) while for HCN the CN force constant is 18.8 mdyn Å^{-1} (see Sect. 2.10). Purcell's calculations (*5*) indicate that the main strengthening of the CN bond in HCN over that of CN^- arises from increased *s* character in the CN σ bond.

The bonding can also be described from a valence bond resonance viewpoint. Figure 4.1-2 depicts the two predominant resonance structures contributing to the CN^- bond structure. When an *X*-*C* sigma bond is formed, using the carbon lone pair, the negative charge is removed and structure (a) is favored, leading to a shorter and stronger CN bond.

For HCN, no additional bonds can be formed; however, for metal-CN bonding there are possibilities for metal-ligand pi bonding which must be considered also. These will be discussed in subsequent sections.

4.2. Linear Dicyanide Complexes, NC*M*CN

Three linear dicyanide complexes are known; $Ag(CN)_2^-$, $Au(CN)_2^-$, and $Hg(CN)_2$. Crystals of $KAg(CN)_2$ (*7*) and $KAu(CN)_2$ (*8*) have been

Fig. 4.1-2 Valence bond resonance structures of CN^-.

shown to contain linear NC*M*CN groups, though it is not conclusive from the crystal structure data that the arrangement cannot be CN*M*NC. The crystal structure by X-ray diffraction and neutron diffraction of $Hg(CN)_2$ suggests a slight deviation from linearity (*9*); however, it is no doubt linear in solution where the crystal forces are absent. The neutron diffraction studies (*9*) do agree with the idea that the CN groups are bonded to the mercury through the carbon atoms. In treating the vibrational spectra of these complexes we have assumed a linear, symmetric structure, NC*M*CN, of $D_{\infty h}$ symmetry. The possibility of the N-bonded structure, CN*M*NC, is considered also.

The NC*M*CN model of $D_{\infty h}$ point group will have seven vibrational frequencies as indicated in Table 4.2-1.

Observed values, as derived from infrared and Raman studies (*3*, *10–12*) are given in Table 4.2-2. When approximate anharmonic corrections (*3*) are applied for the CN stretching vibrations of $Au(CN)_2^-$ and $Hg(CN)_2$ we arrive at the harmonic frequencies of Table 4.2-3. The corrections used were $X_{11} \approx X_{33} \approx -8\ \text{cm}^{-1}$ and $X_{13} \approx -24\ \text{cm}^{-1}$, for $Au(CN)_2^-$ (*3*) and $X_{11} \approx X_{33} \approx -7\ \text{cm}^{-1}$ and $X_{13} \approx -21\ \text{cm}^{-1}$ for $Hg(CN)_2$ (*10*). From Eq. (1.3-1) it is found that, with neglect of $X_{\text{CN},MC}$,

$$\omega_1 = \nu_1 - 2X_{11} - \tfrac{1}{2}X_{13}$$

and

$$\omega_3 = \nu_3 - 2X_{33} - \tfrac{1}{2}X_{13}$$

TABLE 4.2-1

FUNDAMENTAL VIBRATIONS OF NC*M*CN, $D_{\infty h}$

Representation		Activity[a]	Description				
Σ_g^+	ν_1	R	←O	O→	O	←O	O→
	ν_2	R	←O	←O	O	O→	O→
Σ_u^+	ν_3	I	←O	O→	O	O→	←O
	ν_4	I	←O	←O	O→	←O	←O
π	ν_5	R	O↓	↑O	O	O↓	↑O
π_u	ν_6	I	O↓	↑O	↑O	↑O	O↓
	ν_7	I	O↓	O	↑O	O	O↓

[a] R = Raman active; I = infrared active.

TABLE 4.2–2

FREQUENCIES[a] FOR LINEAR NC*M*CN

	$Au(CN)_2^-$	$Hg(CN)_2$	$Ag(CN)_2^-$
ν_1	2164.00 ± 0.40[b]	2197.4[b]	2141[b]
ν_2	448 ± 2[b]	412[b]	~360[c]
ν_3	2145.7 ± 0.2[b]	2197.2[c]	2135[b]
ν_4	436 ± 2[b]	442[c]	390[c]
ν_5	304 ± 2[b]	276[b]	250[b]
ν_6	368 ± 5[c]	341[c]	310[c]
ν_7	100 ± 10[c]	—	107[c]
$\nu_1(^{15}N)$[d]	2133.64 ± 0.35[c]	2166.5[b]	
$\nu_3(^{15}N)$[d]	2115.30 ± 0.08[b]	2166.6[c]	
$\nu_1(^{13}C)$[d]	2115.64 ± 0.30[b]	2148.3[b]	
$\nu_3(^{13}C)$[d]	2099.33 ± 0.38[b]	2149.0[c]	

[a] Values are in cm^{-1}.
[b] Measured in aqueous solution.
[c] From solid state studies (mineral oil mulls).
[d] ^{15}N stands for $M(C^{15}N)_2$ and ^{13}C for $M(^{13}CN)_2$.

There is not much evidence for anharmonic corrections for *M*-C stretching vibrations; however, from the few combinations observed, such corrections appear to be small.

The relations between frequencies and force constants are given below

$$\Sigma_g^+ \quad \lambda_1 + \lambda_2 = (\mu_N + \mu_C)F_{11} + \mu_C F_{22} - 2\mu_C F_{12}$$

$$\lambda_1\lambda_2 = \mu_N\mu_C[F_{11}F_{22} - F_{12}^2]$$

$$\Sigma_u^+ \quad \lambda_3 + \lambda_4 = (\mu_N + \mu_C)F_{33} + (\mu_C + 2\mu_M)F_{44} - 2\mu_C F_{34}$$

$$\lambda_3\lambda_4 = (\mu_N\mu_C + 2\mu_N\mu_M + 2\mu_C\mu_M)(F_{33}F_{44} - F_{34}^2)$$

$$\pi_g \quad \lambda_5 = [\mu_N\rho_{CN}^2 + \mu_C(\rho_{CN} + \rho_{MC})^2]F_{55}$$

$$\pi_u \quad \lambda_6 + \lambda_7 = [\mu_N\rho_{CN}^2 + \mu_C(\rho_{CN} + \rho_{MC})^2 + 2\mu_M\rho_{MC}^2]F_{66} + 2(\mu_C + 2\mu_M)\rho_{MC}^2F_{77} - 2\sqrt{2}\,[\mu_C\rho_{MC}(\rho_{CN} + \rho_{MC}) + 2\mu_M\rho_{MC}^2]F_{67}$$

$$\lambda_6\lambda_7 = 2(\mu_C\mu_N + 2\mu_N\mu_M + 2\mu_C\mu_M)(\rho_{CN}\rho_{MC})^2(F_{66}F_{77} - F_{67}^2)$$

$$\lambda_i = 4\pi^2\nu_i^2c^2, \qquad \nu_i \text{ in cm}^{-1}$$

$$\mu_i = \text{reciprocal of mass of atom } i$$

$$\rho_{ij} = \text{reciprocal of distance } ij$$

The symmetry force constants in these secular equations are defined as

$$F_{11} = F_{CN} + F_{CN,C'N'} \qquad F_{55} = F_\alpha - F_{\alpha\alpha'}$$

$$F_{22} = F_{MC} + F_{MC,MC'} \qquad F_{66} = F_\alpha + F_{\alpha\alpha'}$$

$$F_{12} = F_{CN,MC} + F_{CN,MC'} \qquad F_{77} = F_\beta$$

$$F_{33} = F_{CN} - F_{CN,C'N'} \qquad F_{67} = \sqrt{2}\,F_{\alpha\beta}$$

$$F_{44} = F_{MC} - F_{MC,MC'}$$

$$F_{34} = F_{CN,MC} - F_{CN,MC'}$$

First, let us consider the four stretching vibrations. We can treat these without any knowledge of the bending vibrations as they fall in a different symmetry class. There is no stretch–bend interaction unless we consider cubic and higher order terms in the potential function. Such terms are negligible for small displacements. From the above equations, we see that there are six stretching force constants to calculate from only four stretching frequencies. Therefore, the problem is indeterminate without other data. The frequencies of the ^{13}C and ^{15}N species, as given in Table

TABLE 4.2-3

HARMONIC FREQUENCIES[a] FOR $Au(CN)_2^-$ AND $Hg(CN)_2$

	$Au(CN)_2^-$	$Hg(CN)_2$
ω_1	2192.0	2221.9
ω_3	2173.7	2221.7
$\omega_1(^{15}N)^b$	2160.9	2190.3
$\omega_3(^{15}N)^b$	2142.5	2190.4
$\omega_1(^{13}C)^b$	2142.4	2171.7
$\omega_3(^{13}C)^b$	2126.1	2172.5

[a] Values are in cm^{-1}.
[b] ^{15}N stands for $M(C^{15}N)_2$ and ^{13}C for $M(^{13}CN)_2$.

4.2-2 and Table 4.2-3, have been used to solve the problem for $Au(CN)_2$ (*3*) and $Hg(CN)_2$ (*10*). For $Ag(CN)_2^-$ we do not yet have sufficient information. The results are given in Table 4.2-4.

From these force constants, the interaction displacement coordinates can be calculated, using Eqs. (1.10-4). The results are given in Table 4.2-5.

Several interesting observations can be made on this data. First, the fact that $F_{CN,MC'}$ and $F_{CN,C'N'}$ are close to zero for both $Au(CN)_2^-$ and $Hg(CN)_2$ suggests that they can be neglected for all metal-cyanide complexes. Of course, further substantiating data is desirable. There are some significant differences between $Au(CN)_2^-$ and $Hg(CN)_2$ which indicate some differences in the electronic structure of the two complexes.

TABLE 4.2-4

VALENCE STRETCHING FORCE CONSTANTS[a] FOR $Au(CN)_2^-$ AND $Hg(CN)_2$

	$Au(CN)_2^-$	$Hg(CN)_2$
F_{CN}	17.64 ± 0.18	18.07 ± 0.2
F_{MC}	2.78 ± 0.03	2.59 ± 0.03
$F_{CN,CN'}$	0.03 ± 0.18	0.0 ± 0.2
$F_{MC,MC'}$	0.416 ± 0.025	0.12 ± 0.02
$F_{CN,MC}$	0.3 ± 0.15	0.05 ± 0.15
$F_{CN,MC'}$	0.0 ± 0.15	0.0 ± 0.15

[a] Values are in mdyn $Å^{-1}$.

The primary CN stretching force constant is somewhat smaller for the gold compound. Also the interaction displacement coordinates $(MC)_{MC'}$ and $(MC)_{CN}$ are significantly larger for the gold compound. Thus, stretching a CN bond a small amount would lead to 11% contraction in the adjoining AuC but only 2% contraction in an adjoining HgC bond *for minimum energy.*

Schematics of the structures of the two species are shown in Fig. 4.2-1.

TABLE 4.2-5

INTERACTION DISPLACEMENT COORDINATES FOR $Au(CN)_2^-$ AND $Hg(CN)_2$

	$Au(CN)_2^-$	$Hg(CN)_2$
$(CN)_{C'N'}$[a]	0	0
$(MC)_{MC'}$	−0.15	−0.05
$(CN)_{MC}$	−0.02	0
$(CN)_{MC'}$	0	0
$(MC)_{CN}$	−0.11	−0.02
$(MC)_{C'N'}$	+0.02	0

[a] The meaning of these coordinates is illustrated by the statement that $(MC)_{MC'} = -0.15$ means that a small stretch of a *MC* bond leads to a contraction of 15% of that amount on the other *MC* bond, *for minimum energy.*

The CN groups are believed to form strong metal-carbon sigma bonds. This leads to a valence bond configuration as shown in Fig. 4.2-1(a). For the mercury compound, the atoms now all have a formal neutral charge; however, for $Au(CN)_2^-$ this places a negative charge on the gold atom. The gold atom is not very electronegative and thus attempts to become neutral by donating $d\pi$ electrons to antibonding π orbitals of the CN groups, as indicated in Fig. 4.2-1(b). This is commonly called "back bonding." Placing electrons in antibonding CN orbitals weakens the CN bonds thus accounting for $Au(CN)_2^-$ having a weaker CN bond than $Hg(CN)_2$.

Now let us see how the interaction displacement coordinates fit the picture. If we stretch a CN bond of $Au(CN)_2^-$, we lower the energy of the antibonding CN π orbitals and make them more available for pi bonding. This strengthens and contracts the AuC bond in agreement with the negative value for the $(MC)_{CN}$ interaction displacement coordinate.

Fig. 4.2-1 Schematic valence bond structures of $Hg(CN)_2$ and $Au(CN)_2^-$.

The fact that $(MC)_{CN}$ is much less for $Hg(CN)_2$ than for $Au(CN)_2^-$ is in agreement with the idea that the Hg–CN pi bonding is much less, as discussed above. We expect a larger $(MC)_{MC'}$ interaction for the gold compound, also, as stretching one MC bond weakens its pi bonding, tending to increase the pi bonding in the other MC bond.

This above discussion illustrates how the force constants and interaction displacement coordinates can be correlated with electronic structure.

For most other cyanide complexes, anharmonicity corrections are not known. Therefore, for comparison of CN stretching force constants among different cyanide complexes, one would be more consistent to use the force constants uncorrected for anharmonicity. The only significant change from Table 4.2-4 would be $F_{CN}[Au(CN)_2^-] = 17.17$ and $F_{CN}[Hg(CN)_2] = 17.66$ mdyn Å^{-1}. Thus, the CN stretching force constants of these complexes can be compared with those of CN^- and HCN on two bases, the harmonic frequencies and the observed or anharmonic frequencies. We shall call these results the harmonic and anharmonic force constants. They are given in Table 4.2-6.

The last column of Table 4.2-6 is probably the most appropriate to use for comparison with other metal-cyanide complexes, for which anharmonicity corrections are not known. The force constants given in the table probably cover the range of all cyanide complexes.

TABLE 4.2-6

COMPARISON OF SOME CN STRETCHING FORCE CONSTANTS

	Harmonic[a]	Anharmonic[b]
CN	16.3 mdyn Å^{-1}	15.9
CN^-	16.9	16.5
HCN	18.8	18.2
$Au(CN)_2^-$	17.6	17.2
$Hg(CN)_2$	18.1	17.7

[a] Based on CN stretching frequencies corrected for anharmonicity.

[b] Based on observed CN stretching frequencies.

4.3. Metal Hexacyanides

Quite a number of metal–hexacyanide complexes have been made. They form an interesting group as one can observe trends in various properties as a function of the central metal atom. The vibrational spectra are still rather incomplete though most of the fundamental frequencies have been assigned (*13*) for the three complexes $Co(CN)_6^{-3}$, $Rh(CN)_6^{-3}$,

TABLE 4.3-1

Anharmonic[a] Stretching Force Constants for $K_3Co(CN)_6$, $K_3Rh(CN)_6$, and $K_3Ir(CN)_6$

(mdyn Å^{-1})	$K_3Co(CN)_6$[b]	$K_3Rh(CN)_6$	$K_3Ir(CN)_6$
F_{CN}	17.0 ± 0.2	17.1	16.9
F_{MC}	2.0 ± 0.1	2.3	2.6
$F_{CN,MC}$	0.2 ± 0.1	0.2	0.2
$F_{CN,MC'}$	0 ± 0.1	0	0
$F_{MC,MC'}$	0.45 ± 0.1	0.75	0.69
$F_{MC,MC''}$	0.04	0.03	0.05
$F_{CN,CN'}$	0	0	0
$F_{CN,CN''}$	0	0	0

[a] Calculated from observed frequencies, uncorrected for anharmonicity.

[b] The limits of error given for $K_3Co(CN)_6$ apply also to the Rh and Ir compounds. They are based on the isotopic frequency shift for the cobalt complex and do not include other possible sources of error such as the possibility of large stretch–bend interactions (which have necessarily been neglected). A single prime represents a trans atom, a double prime a cis.

and $Ir(CN)_6^{-3}$. In only one case has an isotopically enriched species been studied; from Raman spectra of $Co(C^{15}N)_6^{-3}$ in aqueous solution, it was concluded (*14*) that the interaction constant $F_{CN,MC}$ is equal to 0.2 ± 0.1 mdyn Å^{-1}. If we assume that this value applies to the rhodium and iridium compounds also, along with the data of (*13*), we find the results given in Table 4.3-1.

The values for F_{CN} in Table 4.3-1 should be compared with the last column of Table 4.2-6. The other force constants should be compared with Table 4.2-4. The CN stretching force constants for these three hexacyanides are seen to be about the same. They are very slightly less than that for $Au(CN)_2^-$. This suggests that the *M*–CN pi bonding is about the same for the hexacyanides as for $Au(CN)_2^-$.

There is a definite increase in the *M*C stretching force constant as we go down the periodic table: $F_{CoC} < F_{RhC} < F_{IrC}$. This implies stronger *M*C sigma bonding, or pi bonding, or both, for the heavier metals. Such an ordering agrees with the usual increase (*15*, *16*) in ligand field Δ from the first transition series to the second and, again, to the third.

A number of hexacyanides have been studied by Nakagawa and Shimanouchi (*17*). They have calculated force constants based on a

TABLE 4.3-2

FUNDAMENTAL VIBRATIONAL FREQUENCIES OF $K_3Ir(CN)_6$ AND $K_4Os(CN)_6$

		$K_3Ir(CN)_6$[a]	$K_4Os(CN)_6$[b]
A_{1g}	ν_1	2170	2109
	ν_2	469	465
E_g	ν_3	2146	2062
	ν_4	450	450
F_{1g}	ν_5	415	
F_{1u}	ν_6	2131	2036
	ν_7	520	550
	ν_8	398	390
	ν_9	—	—
F_{2g}	ν_{10}	483	285?
	ν_{11}	95	100
F_{2u}	ν_{12}	445	—
	ν_{13}	—	—

[a] From (*13*).
[b] From (*18*).

Urey–Bradley type potential field, modified by inclusion of valence type *MC*,*MC*′ and *M*C,CN interactions. Unfortunately the frequency assignments for infrared active *M*C stretching and *M*CN bending vibrations should be reversed in this author's opinion, because their results give a fairly significant negative value to $F_{MC,MC'}$, which should be positive. Therefore their force constants are not reproduced here. A fairly complete vibrational assignment has been reported (*18*) for $Os(CN)_6^{-4}$. It is interesting to compare the results with those for isoelectronic $Ir(CN)_6^{-3}$. Though force constants based on these frequencies have not been calculated for the osmium compound, some interesting facts emerge from Table 4.3-2. The metal carbon bond strengths are about the same, as evidenced by the similar values for the *M*C stretching vibrations (ν_2, ν_4, and ν_8). The CN stretching force constants are significantly less for the osmium complex. This is expected because the greater negative charge imposed on the osmium atom increases the tendency for back bonding which weakens the CN bond, as discussed in Sect. 4.2. The similarity in *MC* frequencies then suggests greater *MC* sigma bonding for $Ir(CN)_6^{-3}$.

For the most part, the bending frequencies are quite uncertain though it appears from (*13*) that the *MCN* bending force constant is in the neighborhood of 0.5 mdyn Å rad^{-2} while the C*M*C bending force constant is considerably smaller, apparently somewhat less than 0.3 mdyn Å rad^{-2}.*

There is still much data lacking for a comprehensive treatment of the hexacyanides. Studies of complexes highly enriched in ^{13}C or ^{15}N should

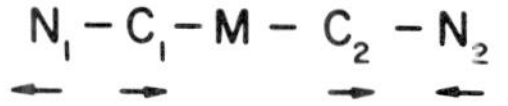

Fig. 4.3-1 Antisymmetric stretching mode of $M(CN)_2$.

be quite important in determining the *MC*,CN interaction constants and in confirming the proper frequency assignment.

Infrared studies (*19*) of intensities of CN stretching vibrations have yielded interesting information. The infrared-active CN stretching frequency of the hexacyanides was found to vary widely among the different complexes. A schematic representation of this normal vibration is shown in Fig. 4.3-1. As one CN expands, the other contracts. There are lesser changes in C_1M (contracts) and C_2M (expands). The other four C_iN_i and corresponding C_iM groups can be neglected as they will not change in first order for this member of the triply degenerate set of ν_6.

The integrated absorption coefficient k_6 is given in Eq. (4.3-1) as a function of the change in dipole moment during the vibration.

$$k_6 = \alpha(\partial\mu/\partial Q_i)^2 \qquad 4.3\text{-}1$$

α is a proportionality constant, and $\partial\mu/\partial Q_i$ is the change in dipole moment accompanying the characteristic oscillation of normal coordinate Q.

In Table 4.3-3 are presented the results (*19*) of studies on some hexacyanides in aqueous solution. Note that the intensity increases rather strikingly with increase in number of d_ϵ electrons. This is highly suggestive that these electrons are involved in the bonding changes which accompany the vibrational displacements. Since the d_ϵ orbitals are π orbitals this suggests metal–ligand pi bonding (so-called backbonding). Thus, as in Sect. 4.2, we hypothesize that the six cyanide–metal sigma bonds

* However, recent studies by the author indicate that the C*M*C bending constant is as large as the *M*CN bending constant.

impose a negative charge on the metal, which then depletes this charge by partial back donation of its d_ϵ electrons to the CN antibonding pi orbitals. Now as C_1N_1 stretches, the C_1N_1 antibonding pi orbitals become more available for accepting the d_ϵ electrons of the metal. At the same time, C_2N_2 is contracting, thus making its antibonding pi orbitals less available for electrons, resulting in decreased MC_2 pi bonding. Therefore, during the phase of the vibration ν_6 represented in Fig. 4.3-1, there is a net flow of d_ϵ electrons in the opposite direction. This shift of electrons back and forth during the vibration leads to a change in

TABLE 4.3-3

VIBRATIONAL FREQUENCY AND INTENSITY DATA FOR SOME METAL HEXACYANIDES

Solution	d_ε	ν_6 (cm^{-1})	ν_7 (cm^{-1})	ν_8 (cm^{-1})	$k_6 \times 10^{-3}$ (M l cm^{-2})
$K_3Cr(CN)_6$	3	2128	458	339	2.1
$K_3Mn(CN)_6$	4	2112	483	361	8.2
$K_3Fe(CN)_6$	5	2118	506	389	12.3
$K_3Co(CN)_6$	6	2129	564	416	18.3
$K_4Fe(CN)_6$	6	2044	583	416	92.0

dipole moment of the ion oscillating with the frequency of the vibration ν_6. Thus, the intensity of ν_6 serves as a measure of the extent of metal-cyanide pi bonding. With more d_ϵ electrons to donate the metal $\rightarrow$ cyanide pi bonding should be stronger, the intensity of ν_6 greater, and the metal carbon bond stronger. Though metal–carbon force constants have not been calculated for the complexes listed in Table 4.3-3, the fact that the MC stretching frequency ν_8 increases significantly with increase in d_ϵ electrons indicates that the MC force constant follows the same trend.

The much greater intensity for the Fe II complex over that of isoelectronic $Co(CN)_6^{-3}$ arises because of the greater negative charge on the Fe atom without back donation.

The similar values of ν_6 in Table 4.3-3 for the M III complexes indicates that the CN force constant does not change appreciably. Thus the increased metal–CN pi bonding which is expected to lower the CN force constant has not done so. This leads to the conclusion that the increased $M \rightarrow$ CN pi bonding which tends to lower F_{CN} is accompanied by an increase in M–C sigma bonding which tends to raise F_{CN}. This is analogous to the explanation (*16*) given for the similar ligand field Δ for $Fe(CN)_6^{-4}$ and $Fe(CN)_6^{-3}$.

4.4. Square Planar Metal Tetracyanides

The tetracyano-complexes of Ni(II), Pd(II), Pt(II), and Au(III) have the D_{4h} square planar structure. A detailed normal coordinate treatment of this general class of molecule is given in Appendix II. The spectra of these four compounds have been studied; however, many of the fundamentals have not been observed or assigned. For D_{4h} $M(CN)_4$, as shown in Appendix II, there are sixteen fundamental vibrations

$$[2A_{1g} + A_{2g} + 2B_{1g} + 2B_{2g} + E_g + 2A_{2u} + 2B_{2u} + 4E_u].$$

The A_{1g} frequencies are assigned from Raman spectra and the E_u frequencies from infrared spectra, as in Table 4.4-1. Some of the values in Table 4.4-1 are for aqueous solution but most apply to the solid. Where both have been observed, the solid–solution shift is not great. Only for

TABLE 4.4-1

Fundamental Vibrational Frequencies of Square–Planar Tetracyanides

		$Ni(CN)_4^{-2}$ [a]	$Pd(CN)_4^{-2}$ [b]	$Pt(CN)_4^{-2}$ [c]	$Au(CN)_4^{-}$ [d]
A_{1g}	ν_1	2149	2161	2168	2207
	ν_2	419	430	465	459
A_{2g}	ν_3	325			
B_{1g}	ν_4	2141	2147	2149	2198
	ν_5	405		455	450
B_{2g}	ν_6	488		318	422
	ν_7	—		95	110
E_u	ν_8	2130	2142[e]	2150	2189
	ν_9	(543) [f]	(503) [e,f]	(505) [f]	462
	ν_{10}	(427)	(393)	(411, 300)	415
	ν_{11}	—			—
A_{2u}	ν_{12}	448			—
	ν_{13}	—			—
B_{2u}	ν_{14}	—			—
	ν_{15}	54?			—
E_g	ν_{16}	280	295		298

[a] From (*20*).
[b] From (*21*).
[c] From (*22*).
[d] From (*23*), (*24*).
[e] From (*25*).
[f] These assignments are in doubt.

$Au(CN)_4^-$ has a systematic study been made (*23*, *24*). It is highly desirable for studies of the completely substituted ^{13}C and ^{15}N species of these square–planar complexes to be carried out so that we have a measure of potential constants.

Force constants are not given here for the D_{4h} tetracyanides, because the data is insufficient for conclusive results. However, two observations can can be made from the A_{1g} and E_g frequencies. From the symmetric *M*C stretching frequency ν_2, it is apparent that the metal–carbon bond becomes stronger as we go down the periodic table; $F_{NiC} < F_{PdC} < F_{PtC}$. This is, of course, analogous to the discussion in the previous section for the hexacyanides of Co, Rh, and Ir. It agrees with the usual increase (*15*, *16*) in ligand field Δ from the first transition series to the second and, again, to the third.

The CN force constant for $Au(CN)_4^-$ is apparently greater than that for $Pt(CN)_4^{-2}$, though the *M*C stretching force constants are similar in magnitude. This is in line with the argument (*19*) that the greater positive charge on the Au III atom leads to strong *M*–C sigma bonding and less *M*–CN pi bonding.

4.5. Tetrahedral Metal Tetracyanides

In contrast to the square planar $M(CN)_4$ ions, which have sixteen fundamental vibrations, the tetrahedral $M(CN)_4$ ions have only nine fundamental vibrations $[2A_1 + 2E + 4F_2 + F_1]$. These have been studied for the tetracyano complexes of Zn (II), Cd (II), Hg (II), and Cu (I). The results are accumulated in Table 4.5-1. Approximate stretching force constants are given in Table 4.5-2. These are calculated with neglect of *M*C, CN interactions. Such interactions are not known at present, though from the work on $KAu(CN)_2$ and $Hg(CN)_2$, as discussed in Sect. 4.2, it is probable that $F_{MC,CN}$ is between zero and +0.3 mdyn Å^{-1}, while $F_{MC,C'N'}$ should be close to zero. This means that the "anharmonic" F_{CN} may be slightly lower than given in the table. It is apparent, however, that the CN force constant of $Cu(CN)_4^{-3}$ is definitely smaller than for the other three tetrahedral complexes listed. This suggests that there is significantly greater metal–ligand pi bonding for the Cu (I) complex than for the others. Thus, as discussed in the previous two sections, we expect the lower valent metal ion to form weaker *M*–C sigma bonds and stronger *M*–CN pi bonds.

The metal–carbon bonds are quite a bit weaker than for the square planar ions; thus, the primary stretching force constants of $Au(CN)_4^-$

TABLE 4.5-1

FUNDAMENTAL FREQUENCIES FOR TETRAHEDRAL $M(CN)_4$ IONS

		$Zn(CN)_4^{-2}$ [a]	$Cd(CN)_4^{-2}$ [a]	$Hg(CN)_4^{-2}$ [a]	$Cu(CN)_4^{-3}$ [b]
A_1	ν_1	2157	2149	2149	2094
	ν_2	342	324	335	(288)[c]
E	ν_3	—	—	—	324
	ν_4	—	—	—	(63)[d]
F_2	ν_5	2152	2145	2146	2079
	ν_6	359	316	330	364
	ν_7	315	250	235	306
	ν_8	(71)[d]	(61)[d]	(54)[d]	(74)[d]
F_1	ν_9	(230)[d]	(194)[d]	(180)[d]	(250)[c]

[a] From (*26*).
[b] From (*27*).
[c] This assignment is doubtful in view of the work by Poulet and Mathieu (*28*).
[d] Assigned from combination bands.

are (*24*) $F_{CN} = 17.5$ and $F_{AuC} = 2.8$ mdyn Å^{-1}. The AuC bond in $Au(CN)_4^-$ is seen to be considerably stronger than those of the tetrahedral tetracyanides. Of course, this is not a satisfactory comparison as the metal valence states are not the same, and we expect the trivalent Au atom to form stronger *M*C sigma bonds than Zn, Cd, and Hg. Unfortunately, we do not have a direct force constant comparison for the divalent square planar and tetrahedral complexes as some of the frequency assignments are uncertain. However, if we compare the symmetric *M*C stretching frequencies ν_2 for Ni, Pd, and Pt (Table 4.4-1) with those for Zn, Cd, and Hg (Table 4.5-1), we see that the *M*C bonds are apparently significantly stronger for the square planar complexes. This may be attributed to the ligand field stabilization for the d^8 complexes [Ni (II), Pd (II), and Pt (II)].

TABLE 4.5-2

STRETCHING FORCE CONSTANTS FOR SOME TETRACYANIDES OF T_d SYMMETRY

Force constants[a]	$Zn(CN)_4^{-2}$	$Cd(CN)_4^{-2}$	$Hg(CN)_4^{-2}$	$Cu(CN)_4^{-3}$
F_{CN}	17.2	17.1	17.1	16.2
F_{MC}	1.3	1.3	1.5	1.3?
$F_{MCMC'}$	0.2	0.1	0.1	?

[a] Units are mdyn Å^{-1}. These are "anharmonic" force constants.

One other probable tetrahedral cyanide complex should be mentioned: namely, $Ni(CN)_4^{-4}$. It is not very stable but El Sayed and Sheline *(29)* have observed the CN stretching vibration for solid $K_4Ni(CN)_4$ at 1985 cm^{-1}. This would lead to a CN stretching force constant of the order of 14.5 to 15 mdyn Å^{-1}, indicating very strong metal–cyanide pi bonding (probably accompanied by quite weak metal–carbon sigma bonding). A thorough study of this complex should prove quite interesting.

4.6. Mixed Cyanide Halide Complexes

Some interesting mixed metal–cyanide–halide complexes have been prepared. Fairly complete vibrational spectra and force constants have been reported for $Au(CN)_2X_2^-$ *(30, 31)* and $Pt(CN)_4X_2^{-2}$ *(32)* with $X = $ Cl, Br, and I. The compounds were prepared by oxidation of $Au(CN)_2^-$ or $Pt(CN)_4^{-2}$ with the free halogen molecule. The question arises, are the halides cis or trans in the resulting complexes? If they are trans, the resulting symmetry is D_{2h} for $Au(CN)_2X_2^-$ and D_{4h} for $Pt(CN)_4X_2^{-2}$. However, if the X atoms are cis to each other, the symmetries are C_{2v} for both types. The selection rules for the CN stretching frequencies are summarized in Table 4.6-1. Thus, for $Au(CN)_2Cl_2^-$, the cis compound should have two vibrations, each active in Raman and infrared. The intensity of absorption in the infrared should be about the same for both vibrations.

In Fig. 4.6-1 are shown schematic diagrams of the two vibrations. The changing dipole moment in each MCN group should be about the same for the two vibrations. They will add vectorially to give the component dipole moment change indicated by the dashed arrows in Fig. 4.6-1. Though differing by 90° in orientation, these vectors should have about the same length, so the intensity of absorption should be about the same. There is observed *(30)* only one infrared CN stretching frequency [at 2181 cm^{-1} for $Au(CN)_2Cl_2^-$] and one Raman CN stretching frequency [at 2199 cm^{-1} for $Au(CN)_2Cl_2^-$]. As seen in Table 4.6-1, this is in agreement with the selection rules for the trans structure but not for the cis structure.

Similarly, for $K_2Pt(CN)_4Cl_2$, there were observed *(32)* two Raman active CN stretching frequencies (2196 and 2186 cm^{-1}) but only one infrared active CN stretching frequency (2176 cm^{-1}). This again indicates the trans structure. Actually one would expect the trans isomers to form because this requires no atom rearrangement when the halides are added.

TABLE 4.6-1

SELECTION RULES FOR CN STRETCH OF cis AND trans $Au(CN)_2X_2^-$ AND $Pt(CN)_4X_2^{-2}$

$Au(CN)_2X_2^-$		$Pt(CN)_4X_2^{-2}$	
cis (C_{2v})	trans (D_{2h})	cis (C_{2v})	trans (D_{4h})
A_1, IR,R B_1 IR,R	A_g R B_{1u} IR	$2A_1$ IR,R B_1 IR,R B_2 IR,R	A_{1g} R B_{1g} R E_u IR

The vibrational frequencies were assigned on the basis of the trans structures. An ambiguity arises with most metal–cyanide complexes because the infrared active M–C stretching and MCN bending vibrations are in the same frequency region. Therefore it is generally not possible to make a conclusive assignment of the frequencies without supplementary information. For $KAu(CN)_2Cl_2$ and $KAu(CN)_2Br_2$, such information was obtained by studying the infrared spectrum of the ^{13}C and ^{15}N isotopically enriched species (*31*).

The observed isotope shifts pointed to conclusive frequency assignments. By analogy, this data settles the assignments for the iodide complex and for the $Pt(CN)_4X_2^{-2}$ complexes also. It appears that generally the M–C stretching vibration shows significantly more intense infrared absorption than the MCN bending vibration. This appears to be a good rule to go by, though, without further evidence, it cannot be accepted with 100% confidence.

The frequency assignments for these mixed cyanide–halide complexes are given in Tables 4.6-2 and 4.6-3.

A glance at Tables 4.6-2 and 4.6-3 shows that there are many fundamentals unobserved. This is partly due to the lack of Raman data for

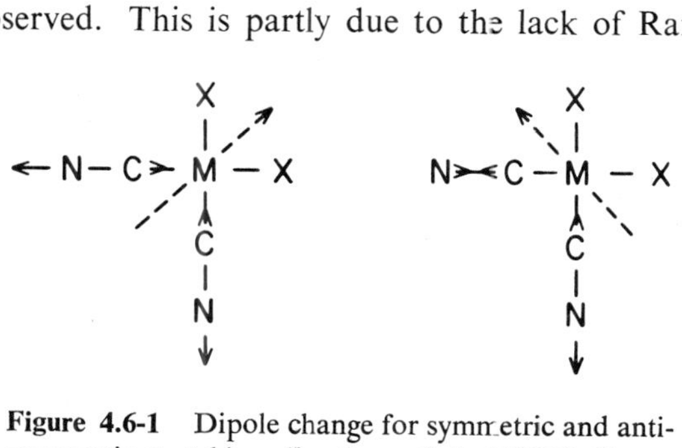

Figure 4.6-1 Dipole change for symmetric and antisymmetric stretching vibrations of cis-$M(CN)_2X_2$.

the highly colored bromides and iodides. But even for the pale yellow chlorides, there are some gaps in frequency assignments. For this reason, a good conclusive force constant solution is not possible. However, one can make some conclusions about the bonding from the observed frequencies. Both for the Au (III) and Pt (IV) complexes, there is a slight

TABLE 4.6-2

FUNDAMENTAL FREQUENCIES[a] FOR $KAu(CN)_2X_2$

Rep[b]	Type[c]	$Au(CN)_2Cl_2^-$	$Au(CN)_2Br_2^-$	$Au(CN)_2I_2^-$
A_{1g}	$\nu_1(\nu_{CN})$	2199	[2199][d]	2186
	$\nu_2(\nu_{MC})$	461	[456][d]	—
	$\nu_3(\nu_{MX})$	342	209	—
B_{1g}	$\nu_4(\delta_{MCN})$	[368][d]	—	[364][d]
	$\nu_5(\delta_{XMC})$	126	—	—
B_{3g}	$\delta_6(\pi_{MCN})$	297	[302][d]	—
B_{1u}	$\nu_7(\pi_{MCN})$	[301][d]	—	—
	$\nu_8(\pi_{CMC})$	—	—	—
	$\nu_9(\pi_{XMX})$	—	—	—
B_{2u}	$\nu_{10}(\nu_{CN})$	2181	2175	2166
	$\nu_{11}(\nu_{MC})$	430	430	430
	$\nu_{12}(\delta_{XMC})$	126	—	—
B_{3u}	$\nu_{13}(\nu_{MX})$	364	258	203
	$\nu_{14}(\delta_{MCN})$	456	455	448
	$\nu_{15}(\delta_{XMC})$	—	—	—

[a] Values are in cm^{-1}.
[b] Representation under D_{2h} point group. All g vibrations are Raman active only; all u vibrations are infrared active only.
[c] ν refers to stretching, δ to "in-plane" bending, and π to "out-of-plane" bending.
[d] Estimated from combination bands.

decrease in CN bond stretch from Cl to Br to I. The metal carbon bonds and *M*CN bending force constants are not appreciably affected by varying the halide ion. The metal–halide frequency shows the expected dependency on halide mass.

By making certain assumptions, force constant solutions have been calculated (*31*, *32*). The most critical assumption is, perhaps, that $F_{MC,CN} \approx$ zero. This was shown to be a good estimate for $Hg(CN)_2$ (see Sect. 4.2) for which the bonding is expected to be somewhat similar due to the high valency of the metal.

For comparison purposes, some of the force constants are given in Table 4.6-4.

TABLE 4.6-3

FUNDAMENTAL FREQUENCIES[a] FOR $K_2Pt(CN)_4X_2$

Rep[b]	Type[c]	$Pt(CN)_4Cl_2^{-2}$	$Pt(CN)_4Br_2^{-2}$	$Pt(CN)_4I_2^{-2}$
A_{1g}	$\nu_1(\nu_{CN})$	2196	2193	2181
	$\nu_2(\nu_{MC})$	463	(467)[d]	(467)[d]
	$\nu_3(\nu_{MX})$	330	—	—
A_{2g}	$\nu_4(\delta_{MCN})$	—	—	—
B_{1g}	$\nu_5(\nu_{CN})$	2186	—	2170
	$\nu_6(\nu_{MC})$	455	—	—
B_{2g}	$\nu_7(\delta_{MCN})$	—	—	—
	$\nu_8(\delta_{CMC})$	—	—	—
E_g	$\nu_9(\delta_{MCN})$	—	—	—
	$\nu_{10}(\delta_{CMX})$	120	—	—
A_{2u}	$\nu_{11}(\nu_{MX})$	349	247	199
	$\nu_{12}(\delta_{MCN})$	496	494	490
	$\nu_{13}(\delta_{CMX})$	—	—	—
B_{2u}	$\nu_{14}(\delta_{MCN})$	—	—	—
	$\nu_{15}(\delta_{MCX})$	—	—	—
E_u	$\nu_{16}(\nu_{CN})$	2176	2172	2161
	$\nu_{17}(\nu_{MC})$	412	412	407
	$\nu_{18}(\delta_{MCN})$	473	476	476
	$\nu_{19}(\delta_{CMC})$	—	—	—
	$\nu_{20}(\delta_{CMX})$	—	—	—

[a] Values are in cm^{-1}.

[b] Representation under point group D_{4h}. Raman active—A_{1g}, B_{1g}, B_{2g}, E_g; infrared active—A_{2u}, E_u; inactive—A_{2g}, B_{2u}.

[c] ν refers to stretching; δ to bending.

[d] Estimated from combination bands.

TABLE 4.6-4

FORCE CONSTANTS[a] OF $Au(CN)_2X_2^-$ AND $Pt(CN)_4X_2^{-2}$

	F_{CN}	F_{MC}	F_{MX}	$F_{MC,MC'}$	$F_{MX,MX'}$	$F_{\beta(MCN)}$
$Au(CN)_2Cl_2^-$	17.4	2.89	2.22	0.57	0.22	0.39
$Au(CN)_2Br_2^-$	17.3	2.85	1.85	0.52	0.21	0.38
$Au(CN)_2I_2^-$	17.2	2.8	1.4	—	—	0.37
$Au(CN)_4^-$	17.5	2.78	—	0.57	—	0.43
$Pt(CN)_4Cl_2^{-2}$	17.3	2.79	2.09	0.63	0.19	0.5
$Pt(CN)_4Br_2^{-2}$	17.2	2.81	1.73	0.65	—	0.5
$Pt(CN)_4I_2^{-2}$	17.1	2.78	1.37	0.68	—	0.5
$PtCl_6^{-2}$			2.0 (1.7 for bromide)			

[a] All units are mdyn $Å^{-1}$ except for F_β which is in mdyn Å rad^{-2}.

The force constants are remarkably similar for the different complexes, though it appears that the metal–halide bond is slightly weaker and the MCN group more resistant to deformation for the Pt complexes. Also the MC bond is a shade weaker, but the MC,MC' interaction is greater for the Pt complexes. The small differences in force constants for the Au and Pt complexes are no doubt related to the difference in nuclear kernel charge on the metal. It would be desirable to have more frequency data, including accurate isotopic shifts, before making much of such subtleties.

The metal halide force constants are about the same as in the pure halide complexes; $F_{\mathrm{AuX}} \approx 2.15$ for $AuCl_4^-$ (*33*), 1.8 for $AuBr_4^-$ (*33*), 2.0 for $PtCl_6^{-2}$, and 1.7 for $PtBr_6^{-2}$. The latter two values are derived from Tables 3.2-4 and 3.2-5 and Eqs. (3.2-6)–(3.2-8). The fact that the spectra and force constants for the mixed cyanide–halide complexes are quite similar to the analogous quantities in the pure cyanides and pure halides shows that cyanides are similar to halides in their metal–bonding properties. It would be most interesting to study the vibrational spectra of the cis isomers of the mixed cyanide–halide complexes and to study the $Au(CN)_3Cl^-$ complexes (*34*).

4.7. Anharmonicity Corrections for Metal Cyanide Complexes

All the force constants reported for metal cyanide complexes, with the exception of F_{CN} for $Au(CN)_2^-$ and $Hg(CN)_2$, have been calculated from the observed frequencies. We call these the "anharmonic" force constants as the frequencies were not corrected for anharmonicity. As discussed in Sect. 4.2, they should be compared with the values in the last column of Table 4.2-6. It is seen from that table that corrections for anharmonicity lead to "harmonic" CN force constants some 0.4–0.6 mdyn Å^{-1} higher than the anharmonic values.

The observed energy levels are related to the harmonic frequencies through Eq. (1.3-1).

$$G_{v_1,v_2,\ldots} = \sum_i \omega_i\left(v_i + \frac{d_i}{2}\right) + \sum_{i \le j} X_{ij}\left(v_i + \frac{d_i}{2}\right)\left(v_j + \frac{d_j}{2}\right) + \cdots \quad (4.7\text{-}1)$$

The ω_i are the harmonic frequencies, d_i is the degeneracy of ω_i, and the X_{ij} are the anharmonicity constants. From observed combination bands with i equal to a CN stretch and j not equal to a CN stretch, it is apparent that these particular X_{ij} values are generally negligible. The X_{ij} with neither i nor j a CN stretch have not been observed in a sufficient number of cases to make a generalization though they appear to be only a few wave numbers. The conclusion is that their neglect does not have a serious effect on the force constant calculation. However, for all cyanide complexes there can be observed infrared active combinations of two cyanide stretching vibrations. From these, anharmonicity constants can be determined which range from 0 to 30 cm^{-1}. It is of interest to examine these values and consider their effect on the force constants. First, let us consider what binary combinations of CN stretching vibrations are available. The CN stretching vibrations for different classes of compounds

TABLE 4.7-1

CLASSIFICATION OF CN STRETCHING VIBRATIONS FOR METAL CYANIDE COMPLEXES

Linear, $M(CN)_2$	$(D_{\infty h})$	ν_s Symmetric Σ_g^+
Planar, trans $M(CN)_2X_2$	(D_{2h})	ν_a Antisymmetric Σ_u^+
Square planar, $M(CN)_4$	(D_{4h})	ν_s Symmetric A_{1g}
		ν_i B_{1g}
		ν_a Antisymmetric E_u
Tetrahedral, $M(CN)_4$	(T_d)	ν_s Symmetric A_1
		ν_a Antisymmetric F_2
Octahedral, $M(CN)_6$	(O_h)	ν_s Symmetric A_{1g}
		ν_i E_g
		ν_a Antisymmetric F_{1u}

are given in Table 4.7-1. For the D_{4h} and O_h molecules, ν_i represents a vibration which is not totally symmetric, but still preserves the center of symmetry. There are several classes of CN,CN interaction constants; $X_{s,s}$, $X_{i,i}$, $X_{a,a}$, $X_{s,i}$, $X_{s,a}$, and $X_{i,a}$. To determine these values we must observe analogous combination bands; thus, we can determine $X_{s,a}$ from the combination band $(\nu_s + \nu_a)$ and the observed fundamentals, ν_s and ν_a. Unfortunately many of the necessary combinations are infrared inactive and combination bands in Raman spectra are generally too weak to observe. Table 4.7-2 lists the anharmonicity constants which have been calculated, classified as in Table 4.7-1.

It is evident that within a given class of metal cyanide complexes, the anharmonicity constants follow a certain pattern. However, from the limited amount of data in Table 4.7-2 it is apparent that it is not yet possible to make generalized predictions about the various interaction constants. There is no data at all for $X_{s,i}$ and $X_{i.i}$ and that for $X_{s,s}$ is very meager. There are several hopes for attaining such data. For example, laser Raman spectrometry may provide sufficient intensity for observation of the Raman active combination bands. Further careful infrared studies of complexes in the crystal state, where the symmetry is lowered, should be helpful. Also an investigation of the ternary combination region (6200–6600 cm^{-1}), as with hexacarbonyls (*36*), may be fruitful. Until the situation is improved, it is more appropriate to neglect anharmonicity

TABLE 4.7-2

ANHARMONIC CONSTANTS[a] FOR METAL CYANIDE COMPLEXES

	$X_{s,a}$	$X_{i,a}$	$X_{a,a}$	$X_{s,s}$
$KAg(CN)_2$	~ −26			
$KAu(CN)_2$	−23.5		−10[b]	−9[b]
$Hg(CN)_2$	−20.5			
$K_2Zn(CN)_4$	−27			
$K_2Cd(CN)_4$	−27			
$K_2Hg(CN)_4$	−27			
$K_3Cu(CN)_4$	−6.3[c]		−12, −14[c] −3	−1.5[c]
$K_2Ni(CN)_4$	0[c]	−17[c]	−12[c]	
$KAu(CN)_4$	−3	−20		
$K_3Co(CN)_6$	−2	−17		
$K_3Rh(CN)_6$	−2	−18		
$K_3Ir(CN)_6$	−4	−17		
$KAu(CN)_2Cl_2$	−26			
$KAu(CN)_2I_2$	−22		−14	
$K_2Pt(CN)_4Cl_2$	−4	−17		
$K_2Pt(CN)_4Br_2$	−4			
$K_2Pt(CN)_4I_2$	0	−15		

[a] Values are in cm^{-1}.

[b] Estimated from spectra of mixed isotopic species (*35*).

[c] These values were estimated from observed combinations made infrared active by lowering of symmetry by the crystal lattice. The assignments (*27*) are not certain.

for comparison among different metal–cyanide complexes, realizing, of course, that relative uncertainties of many wave numbers exist in the fundamental frequencies. Such uncertainties will be reflected particularly in the CN–CN interaction constants.

REFERENCES

1. N. A. Curry and W. A. Runciman, *Acta Cryst.*, **12,** 674 (1959).
2. A. Sequeira and R. Chidambaram, *Acta Cryst.*, **20,** 910 (1966).
3. L. H. Jones, *J. Chem. Phys.*, **43,** 594 (1965).

4. C. J. Ballhausen and H. B. Gray, *Molecular Orbital Theory*, Benjamin, New York, 1964, Chapters 2 and 4.
5. K. F. Purcell, *J. Am. Chem. Soc.*, **89,** 247 (1967).
6. K. F. Purcell and R. S. Drago, *J. Am. Chem. Soc.*, **88,** 919 (1966).
7. J. L. Hoard, *Z. Krist.*, **84,** 231 (1933).
8. A. Rosenzweig and D. T. Cromer, *Acta Cryst.*, **12,** 709 (1959).
9. J. Hvoslef, *Acta Chem. Scand.*, **12,** 1568 (1958).
10. L. H. Jones, *J. Chem. Phys.*, **44,** 3643 (1966).
11. L. H. Jones, *Spectrochim. Acta*, **19,** 1675 (1963).
12. L. H. Jones, *J. Chem. Phys.*, **26,** 1578 (1957).
13. L. H. Jones, *J. Chem. Phys.*, **41,** 856 (1964).
14. Unpublished work by the author.
15. L. E. Orgel, *An Introduction to Transition Metal Chemistry: Ligand Field Theory,* Wiley, New York, 1960, p. 45.
16. H. B. Gray and N. A. Beach, *J. Am. Chem. Soc.*, **85,** 2922 (1963).
17. I. Nakagawa and T. Shimanouchi, *Spectrochim. Acta*, **18,** 101 (1962).
18. J. P. Mathieu and H. Poulet, *Spectrochim. Acta*, **19,** 1967 (1963).
19. L. H. Jones, *Inorg. Chem.*, **2,** 777 (1963).
20. R. L. McCullough, L. H. Jones, and G. A. Crosby, *Spectrochim. Acta*, **16,** 929 (1960).
21. Unpublished work in this laboratory.
22. D. M. Sweeny, I. Nakagawa, S. Mizushima, and J. V. Quagliano, *J. Am. Chem. Soc.* **78,** 889 (1956).
23. L. H. Jones and J. M. Smith, *J. Chem. Phys.*, **41,** 2507 (1964).
24. L. H. Jones, *Inorg. Chem.*, **4,** 1472 (1965).
25. A. Hidalgo and J. P. Mathieu, *Compt. rend.*, **249,** 233 (1959).
26. L. H. Jones, *Spectrochim. Acta*, **17,** 188 (1961).
27. L. H. Jones, *J. Chem. Phys.*, **29,** 463 (1958).
28. A. Poulet and J. P. Mathieu, *Spectrochim. Acta.*, **15,** 932 (1959).
29. M. F. El Sayed and R. K. Sheline, *J. Am. Chem. Soc.* **80,** 2047 (1958).
30. L. H. Jones, *Inorg. Chem.*, **3,** 1581 (1964).
31. L. H. Jones, *Inorg. Chem.*, **4,** 1472 (1965).
32. L. H. Jones and J. M. Smith, *Inorg. Chem.*, **4,** 1677 (1965).
33. L. H. Jones, *Coord, Chem. Rev.*, **1,** 351 (1966).
34. J. M. Smith, L. H. Jones, I. K. Kressin, and R. A. Penneman, *Inorg. Chem.*, **4,** 369 (1965).
35. L. H. Jones, *J. Chem. Phys.*, **27,** 468 (1957).
36. J. M. Smith and L. H. Jones, *J. Mol. Spectry.*, **20,** 248 (1966).

Chapter 5

Metal Carbonyls

Over the years, the metal carbonyls have been extremely interesting to inorganic chemists because of the unusual valence states of the metals in these compounds and the variety which can be prepared. There are the monomeric carbonyls such as $Ni(CO)_4$, $Fe(CO)_5$, $Cr(CO)_6$, etc.; polymers such as $Co_2(CO)_8$, $Mn_2(CO)_{10}$, etc.; and mixed complexes such as $Co(CO)_3NO$, Cp–$Mo(CO)_3$, etc. Whenever a new carbonyl is prepared, one generally records the infrared spectrum in the CO stretching region and attempts to correlate the number of CO stretching bands and their frequencies and intensities with the structure of the compound. Recently it has also become fashionable to make the approximation that CO stretching frequencies can be considered as not seriously affected by the many, much smaller force constants in the molecule. The ensuing grossly simplified normal coordinate treatments have yielded CO force constants which show some consistency of pattern, but which have no theoretical foundation except in gross approximation. In this chapter, I hope to show how good (or bad) such approximations may be by comparison with what little conclusive knowledge we do have. I shall follow a reverse chronology, by presenting first the most recent and complete data and then comparing it with the many earlier approximations. First of all, it is necessary to consider the ligand, CO.

5.1. The CO Molecule

The CO molecule is isoelectronic with N_2 and CN^-. One can represent the ground state as

$$CO:\quad (1s\sigma^b)^2(1s\sigma^*)^2(2\sigma_o)^2(2\sigma^b)^2(2\pi^b)^4(2\sigma_c)^2$$

The highest energy orbital, $2\sigma_c$, is nonbonding, located primarily on the carbon atom, and its electron pair is available for donation in sigma bond formation to metals. Here, analogous to CN^-, it is probable that the sigma and pi bonds are polarized toward the more negative oxygen. As discussed by Moffitt (*1*) when CO loses an electron to become CO^+, the bonds are less polarized and therefore stronger. Thus the harmonic force constants are 19.02 mdyn Å^{-1} for CO and 19.80 mdyn Å^{-1} for CO^+.

The stable metal carbonyls, formed by bonding of CO to the metal through the carbon atom*, involve metals in low valence states, often formally zero. This means that the M–C sigma bond formation tends to impose considerable negative charge on the metal atom. It is thought that to overcome this, the metal donates $d\pi$ electrons to antibonding CO pi orbitals. This is analogous to the discussion for $Au(CN)_2^-$ in Sect. 4.2. However, the CO molecule is neutral and thus will accept electrons much more readily than CN^- will (and also donates electrons less easily). Therefore, metal–ligand "backbonding" is considerably more important for metal carbonyls than for metal cyanide-complexes. The backbonding makes the CO group more negative and thus a better donor of the sigma bonding electrons. In comparing metal carbonyls with metal–cyanide complexes, we might then conclude that for the carbonyls the M–C sigma bonding is weaker but the M–C pi backbonding is greater than for the cyanide complexes.

5.2. The Hexacarbonyls of Chromium, Molybdenum, and Tungsten

The best estimates of force constants for carbonyls are those determined for $Cr(CO)_6$, $Mo(CO)_6$, and $W(CO)_6$ (*2*). These are based on the infrared and Raman spectra of three isotopic species each; $M(^{12}C^{16}O)_6$, $M(^{13}C^{16}O)_6$,

* For $Ni(CO_4)$ it has been shown (*1a*) that the CO is bonded to the Ni through the carbon atom (Ni—C—O). Metal-carbon bonding also prevails for $Fe(CO)_5$ as shown in (*1b*).

and $M(^{12}C^{18}O)_6$. These molecules were studied in the solid state, solution, and vapor. The infrared of the solids is not reported as the CO absorption was badly broadened by lattice coupling, apparently. The Raman of the vapors was not observed. Table 5.2-1 lists the fundamental frequency assignments which have been made for the isotopic molybdenum carbonyls. Many of these were not observed but were determined from combination bands. The analogous data for the chromium and tungsten molecules can be found in (*2*), in which there is a discussion of the frequency assignments.

For the CO stretching frequencies, anharmonicity corrections were applied using the data of Smith and Jones (*3*) and correcting for isotopic shifts. Unfortunately, there is not enough data for determining the anharmonicity constants for the other vibrations. However, the available evidence suggests that such corrections are not more than a few wave numbers.

For the A_{1g}, E_g, and F_{1u} symmetry blocks, force constants have been estimated (*2*), for both solution and vapor, using the computer program, FPERT, developed by Schachtschneider (*4*). The F_{1g}, F_{2g}, and F_{2u} blocks were solved for only the vapor as solution frequencies were not available. In order to converge on a solution for the F_{1u} symmetry block, it was necessary to fix the constants for interaction of CO with MCO angles and of CO with CMC angles because the frequencies are rather insensitive to these force constants. The Coriolis coupling constants, estimated from band contours of ν_9 for $Mo(CO)_6$ and $Cr(CO)_6$, were also used to estimate the best force constant solution. There is a strong correlation between ζ_9 and $F_{MC,\alpha}$ (α = CMC angle). The final set of internal coordinate force constants arrived at are given in Table 5.2-2. Definition of the various interaction constants is given in (*2*). The less obvious definitions are reproduced here in Table 5.2-3.

The force constants of Table 5.2-2 must, at present, be considered the best estimates of force constants for any metal carbonyls. The standard deviations given are not true error limits but serve two purposes. They give us an idea of how well the force constants are known, and they serve to test whether any significant differences exist in the force constants for the three hexacarbonyls and for solution and vapor phases. In discussing these statistics, one other fact must be borne in mind. Most of these force constant estimates are based on the somewhat arbitrary assumption that $F_{CO,\beta}$ and $F_{CO,\alpha}$ are zero. Also for $Cr(CO)_6$, they are based on a fixed value for $F_{MC,\alpha}$ which yields a Coriolis coupling constant ζ_9, which agrees with that estimated from the band contour of ν_9 in the vapor.

TABLE 5.2-1

FUNDAMENTAL FREQUENCIES[a] OF $Mo(^kC^lO)_6$

Assign-	ment	Solid			CCl_4 Solution			Vapor[b]		
		$k, l = 12, 16$	13, 16	12, 18	$k, l = 12, 16$	13, 16	12, 18	$k, l = 12, 16$	13, 16	12, 18
A_{1g}	ν_1	2113.6±0.3[c]	2064.0±0.3	2067.4±0.3	2116.7±0.2	2067.2±0.2	2070.7±0.1	2120.7	2072.7	2074.9
	ω_1	2137.1[d]	2086.4	2089.9	2140.2	2089.6	2093.2	2144.2	2095.1	2097.4
	ν_2	406.8±0.3	400.1±0.3	392.5±0.3	402.2±0.3	396.1±0.5	385.6±0.7	391.2	386.8	379.2
E_g	ν_3	2005.2±0.3	1960.0±0.3	1959.9±0.3	2018.8±0.2	1972.9±0.2	1973.1±0.2	2024.8	1978.7	1979.1
	ω_3	2023.7	1977.6	1977.5	2037.3	1990.5	1990.7	2043.3	1996.3	1996.7
	ν_4	392.2±0.3	385.0±0.6	379.6±0.6	392.0±1	386.0±2	375.5±1	381		
F_{1g}	ν_5	346.0[e]	337.0[e]	341.5[e]				341.6	332.3	337.2
F_{1u}	ν_6				1986.1±0.2	1942.7±0.2	1938.6±0.2	2003.0±0.1	1959.1±0.1	1955.2±0.1
	ω_6				2026.2	1981.1	1976.8	2043.1	1997.5	1993.4
	ν_7				592.8±0.4	576.8±0.4	589.5±0.4	595.6±0.4	580.0±0.4	592.7±0.4
	ν_8				367.0±0.5	362.0±0.5	359.3±0.5	367.2±0.5	362.3±0.5	359.7±0.5
	ν_9	91.8±0.5	91.5±1.0	89.0±0.5	91.0±1[f]			81.6±0.5	81.3±0.5	78.2±0.5
F_{2g}	ν_{10}	476.5±0.5	462.0±0.5	475.1±0.5				477.4	461.4	474.2
	ν_{11}	103.7±0.5	103.0±0.5	98.8±0.5	91.0±1	89.5±1	84.0±1	79.2	79.0	75.2
F_{2u}	ν_{12}	508.5±1		507.6±1				507.2	491.8	503.9
	ν_{13}							60		

[a] Values are in cm^{-1}.

[b] Only the F_{1u} frequencies were observed directly for the vapor. The others (A_{1g}, E_g, F_{1g}, F_{2g}, and F_{2u} for the vapor) were estimated from observed combination bands as described in the text.

[c] The ± limits are estimated observational errors.

[d] The ω_i are harmonic CO stretching frequencies, estimated by applying anharmonic corrections as described in the text.

[e] Estimated from $2\nu_5$.

[f] See (*3*).

TABLE 5.2-2

FORCE CONSTANTS[a] OF METAL HEXACARBONYLS

	$Cr(CO)_6$ Soln.	$Cr(CO)_6$ Vapor	$Mo(CO)_6$ Soln.	$Mo(CO)_6$ Vapor	$W(CO)_6$ Soln.	$W(CO)_6$ Vapor
F_{CO}	17.036 (0.051)[b]	17.246 (0.043)	17.141 (0.035)	17.310 (0.056)	17.015 (0.026)	17.222 (0.039)
F_{MC}	2.095 (0.029)	2.096 (0.032)	1.984 (0.052)	1.906 (0.054)	2.330 (0.042)	2.385 (0.057)
$F^{c}_{CO,C'O'}$	0.169 (0.011)	0.211 (0.029)	0.167 (0.016)	0.215 (0.052)	0.186 (0.017)	0.219 (0.017)
$F^{t}_{CO,C'O'}$	0.076 (0.051)	0.008 (0.043)	0.009 (0.035)	−0.050 (0.056)	0.102 (0.026)	−0.011 (0.039)
$F^{c}_{MC,MC'}$	−0.020 (0.002)	−0.019 (0.003)	0.037 (0.003)	0.031 (0.009)	0.052 (0.003)	0.049 (0.002)
$F^{t}_{MC,MC'}$	0.464 (0.029)	0.417 (0.032)	0.635 (0.052)	0.580 (0.054)	0.600 (0.042)	0.522 (0.057)
$F_{MC,CO}$	0.686 (0.046)	0.696 (0.040)	0.734 (0.039)	0.702 (0.051)	0.837 (0.030)	0.804 (0.043)
$F^{C}_{MC,C'O'}$	−0.094 (0.013)	−0.052 (0.043)	−0.107 (0.014)	−0.076 (0.014)	−0.092 (0.010)	−0.052 (0.024)
$F^{t}_{MC,C'O'}$	0.098 (0.046)	0.112 (0.040)	−0.136 (0.039)	−0.128 (0.051)	0.087 (0.030)	0.137 (0.043)
F_{β}		0.460 (0.050)		0.459 (0.020)		0.474 (0.039)
$F_{\beta\beta'}$		0.077 (0.050)		0.066 (0.020)		0.076 (0.039)
$F_{\beta\beta''}$		0.004 (0.025)		0.024 (0.005)		0.006 (0.028)
$F_{\beta\beta'''}$		−0.027 (0.043)		−0.013 (0.029)		−0.045 (0.028)
$F_{\alpha} - F_{\alpha\alpha''}$		0.63 (0.15)		0.66 (0.09)		0.62 (0.07)
$\bar{F}_{\alpha\alpha'} - F_{\alpha\alpha'''}$		0.14 (0.07)		0.14 (0.05)		0.14 (0.04)
$F_{\alpha} - F_{\alpha\alpha''''}$		0.58 (0.08)		0.50 (0.05)		0.50 (0.08)
$F_{\alpha\beta'}$		−0.11 (0.02)		−0.10 (0.02)		−0.12 (0.02)
$F_{\alpha\beta''}$		−0.02 (0.02)		−0.05 (0.02)		−0.06 (0.02)
$F_{\alpha\beta'''}$		−0.02 (0.02)		−0.01 (0.01)		−0.04 (0.02)
$F_{CO,\beta'}$	[0][c]	[0]	[0]	[0]	[0]	[0]
$F_{MC,\beta'}$	−0.07 (0.02)	−0.07 (0.02)	−0.03 (0.02)	−0.05 (0.02)	−0.03 (0.03)	−0.02 (0.05)
$F_{CO,\alpha'}$	[0]	[0]	[0]	[0]	[0]	[0]
$F_{MC,\alpha'}$	[−0.15]	[−0.15]	−0.17 (0.09)	−0.25 (0.06)	−0.13 (0.11)	−0.04 (0.20)

[a] Units are mdyn $Å^{-1}$ for stretching constants, mdyn Å rad^{-2} for bending constants, and mdyn rad^{-1} for stretch–bend interactions.

[b] Numbers in parentheses are least squares standard deviations.

[c] Values in square brackets were fixed in these calculations; they are constraints in the least squares calculation.

TABLE 5.2-3

DEFINITION OF INTERACTION POTENTIAL CONSTANTS FOR HEXACARBONYLS

1. For stretch–stretch interaction, the superscript c stands for cis and t for trans.	
2. $F'_{\beta\beta}$	β_{12} with β_{62}
$F''_{\beta\beta}$	β_{12} with β_{21}
$F'''_{\beta\beta}$	β_{12} with β_{32}
3. $\bar{F}'_{\alpha\alpha}$	$F'_{\alpha\alpha} - f_\alpha$
f_α	coefficient of linear term, $\Sigma\ \alpha_{ij}$, in potential energy to allow for redundancy
$F'_{\alpha\alpha}$	α_{12} with α_{13}
$F''_{\alpha\alpha}$	α_{12} with α_{46}
$F'''_{\alpha\alpha}$	α_{12} with α_{36}
$F''''_{\alpha\alpha}$	α_{12} with α_{14}
4. $F'_{\alpha\beta}$	α_{12} with β_{12}
$F''_{\alpha\beta}$	α_{12} with β_{62}
$F'''_{\alpha\beta}$	α_{12} with β_{32}
5. $F'_{r\beta}$	r_1 with β_{21}
$F'_{r\alpha}$	r_1 with α_{12}

[a] Coordinates are defined in Fig. 5.2-1.

There are several items of interest in contemplation of the results of Table 5.2-2. For example, the actual values are useful for estimating bond strengths and directional forces. Also we are interested in evaluating any significant differences which exist among these three molecules. Perhaps the most interesting question at this stage of the game is whether these force constant estimates can be transferred to other metal carbonyls

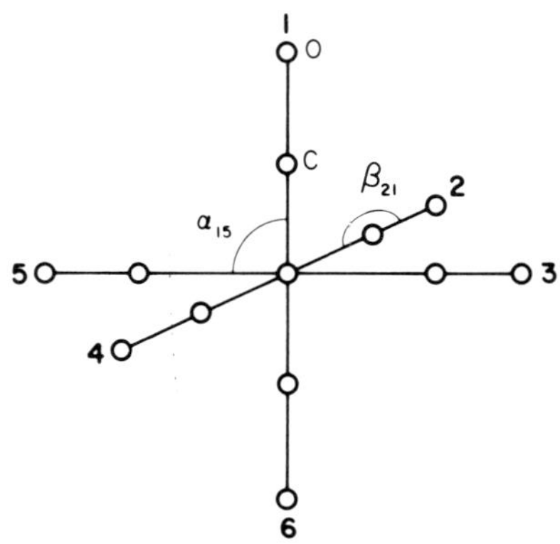

Fig. 5.2-1 Schematic diagram of octahedral hexacarbonyls.

for which we cannot hope to obtain sufficient experimental data to make a satisfactory individual force constant treatment. As pointed out in Sect. 1.13, compliance constants should be more appropriate for transfer among similar types of molecules than are force constants. This is because compliance constants are more unique quantities than force constants, as they do not depend on the coordinate system used to define the potential energy. Therefore, before discussing the force constants, I wish to list the compliance constants estimated from the same vibrational frequencies used to estimate the force constants. These compliance constants are listed in Table 5.2-4. They were determined using the Ottinger–Ryan computer program mentioned in Sect. 1.13.

As with the force constants, it was necessary to fix some of the compliance constants in the F_{1u} symmetry block to achieve convergence to a satisfactory solution. Thus, $C'_{\mathrm{CO},\beta}$ and $C'_{\mathrm{CO},\alpha}$ were fixed at -0.005 and -0.004 rad mdyn^{-1}, respectively. These values were determined from the interaction coordinates, using the approximations: $(\alpha)_{\mathrm{CO}} = (\alpha)_{M\mathrm{C}}(M\mathrm{C})_{\mathrm{CO}}$ and $(\beta)_{\mathrm{CO}} = (\beta)_{M\mathrm{C}}(M\mathrm{C})_{\mathrm{CO}}$. This will be discussed further after the interaction coordinates are evaluated, in the next section. The constant $C'_{M\mathrm{C},\alpha}$ was fixed for $\mathrm{Cr(CO)_6}$ in order to yield a satisfactory value for ζ_9.

The six sets of data (three molecules in solution and vapor) have been subjected to a statistical analysis to look for significant differences. The analyses used were multiple comparison tests using χ^2 at 95% confidence (5). The individual internal coordinate force constants and compliance constants will be discussed in turn.

It is found that a significant difference exists for F_{CO} between solution and vapor. At the 95% confidence level, the mean of F_{CO} for the vapor is greater than the mean for solution. However, at 95% confidence we cannot reject the hypothesis that the solution values have the same mean for the three molecules, and also that the vapor values have the same mean for the three molecules. On the other hand, an analysis of the CO compliance constants of Table 5.2-4 shows not only that the mean of C_{CO} for vapor is significantly less than for solution but that C_{CO} for $\mathrm{W(CO)_6}$ is significantly greater than C_{CO} for $\mathrm{Mo(CO)_6}$ or $\mathrm{Cr(CO)_6}$ in the same phase. Thus, for the CO bond in metal hexacarbonyls, compliance constants are somewhat more definitive than are the force constants. This is also suggested by the much smaller standard deviation, as related to the mean (σ/μ) for C_{CO} than for F_{CO}.

In contrast to the above remarks concerning CO stretching constants, the metal–carbon force constants $F_{M\mathrm{C}}$ are in general determined better than the compliance constants $C_{M\mathrm{C}}$ [except for $\mathrm{W(CO)_6}$ vapor]. In this

TABLE 5.2-4

Compliance Constants of Metal Hexacarbonyls

	$Cr(CO)_6$ Soln.	$Cr(CO)_6$ Vapor	$Mo(CO)_6$ Soln.	$Mo(CO)_6$ Vapor	$W(CO)_6$ Soln.	$W(CO)_6$ Vapor
C_{CO}	0.05975 (0.00004)[b]	0.05900 (0.0008)	0.05971 (0.00005)	0.05897 (0.00013)	0.06017 (0.00004)	0.05928 (0.00006)
C_{MC}	0.568 (0.023)	0.567 (0.037)	0.657 (0.120)	0.800 (0.255)	0.503 (0.064)	0.450 (0.004)
$C^{c}_{CO,C'O'}$	−0.00069 (0.00003)	−0.00077 (0.00007)	−0.00070 (0.00004)	−0.00079 (0.00012)	−0.00081 (0.00004)	−0.00084 (0.00003)
$C^{t}_{CO,C'O'}$	−0.00074 (0.00004)	−0.00051 (0.00008)	−0.00080 (0.00005)	−0.00050 (0.00013)	−0.00091 (0.00004)	−0.00053 (0.00006)
$C^{c}_{MC,MC'}$	0.0023 (0.0003)	0.0026 (0.0007)	−0.0061 (0.0005)	−0.0055 (0.0011)	−0.0069 (0.0003)	−0.0063 (0.0004)
$C^{t}_{MC,MC'}$	−0.173 (0.023)	−0.166 (0.037)	−0.27 (0.12)	−0.39 (0.25)	−0.156 (0.064)	−0.102 (0.004)
$C_{MC,CO}$	−0.0234 (0.0013)	−0.0229 (0.0014)	−0.0290 (0.0019)	−0.0290 (0.0026)	−0.0259 (0.0014)	−0.0221 (0.0008)
$C^{c}_{MC,C'O'}$	0.0022 (0.0002)	0.0013 (0.0006)	0.0023 (0.0003)	0.0015 (0.0009)	0.0025 (0.0002)	0.0019 (0.0003)
$C^{t}_{MC,C'O'}$	0.0098 (0.0013)	0.0093 (0.0014)	0.0150 (0.0019)	0.0152 (0.0026)	0.0101 (0.0014)	0.0082 (0.0008)
C_{β}		2.45 (0.33)		2.59 (0.28)		2.58 (0.20)
C_{α}		1.69 (0.26)		2.05 (0.18)		1.94 (0.16)
$C'_{\beta\beta}$		−0.43 (0.33)		−0.11 (0.28)		−0.10 (0.20)
$C''_{\beta\beta}$		0.11 (0.14)		−0.09 (0.03)		0.05 (0.17)
$C'''_{\beta\beta}$		0.13 (0.30)		0.06 (0.28)		0.37 (0.10)
$C''_{\alpha\alpha}$		−0.67 (0.26)		−0.45 (0.18)		−0.55 (0.16)
$C'_{\alpha\alpha}$		−0.17 (0.12)		−0.18 (0.09)		−0.22 (0.05)
$C'_{\alpha\beta}$		0.48 (0.14)		0.61 (0.12)		0.59 (0.09)
$C''_{\alpha\beta}$		0.01 (0.14)		0.19 (0.12)		0.13 (0.09)
$C'''_{\alpha\beta}$		0.06 (0.14)		−0.04 (0.12)		0.10 (0.07)
$C'_{CO,\beta}$	[−0.0053][c]	[−0.0053]	[−0.0053]	[−0.0053]	[−0.0053]	[−0.0053]
$C'_{MC,\beta}$	0.14 (0.05)	0.14 (0.10)	0.13 (0.07)	0.25 (0.11)	0.10 (0.08)	−0.02 (0.04)
$C'_{CO,\alpha}$	[−0.00375]	[−0.00375]	[−0.00375]	[−0.00375]	[−0.00375]	[−0.00375]
$C'_{MC,\alpha}$	[0.09128]	[0.09128]	0.12 (0.07)	0.21 (0.10)	0.07 (0.05)	0.01 (0.03)

[a] Units are A $mdyn^{-1}$ for stretching constants, $radian^2$ A^{-1} $mdyn^{-1}$ for bending constants, rad $mdyn^{-1}$ for stretch–bend interactions.
[b] Numbers in parentheses are least squares standard deviations.
[c] Values in square brackets were fixed in these calculations; they are constraints in the least squares calculation.

case, the results for F_{MC} are somewhat more definitive than those for C_{MC}. They show that at the 95% confidence level, $W(CO)_6$ has stronger MC bonds than the other two ($F_{MC}^{W} > F_{MC}^{Mo} \approx F_{MC}^{Cr}$, and $C_{MC}^{W} <$ $C_{MC}^{Mo} \approx C_{MC}^{Cr}$). The standard deviations for C_{MC} of $Mo(CO)_6$ are exceedingly large. The reason for this is not apparent.

The only other force constant which shows significant differences at 95% confidence is $F_{MC,MC'}^{c}$. It shows no phase dependence but is definitely greatest for $W(CO)_6$ and least for $Cr(CO)_6$. The analogous compliance constant $C_{MC,MC'}^{c}$ gives a somewhat different statistical analysis. At 95% confidence, $C_{MC,MC'}^{c}$ $[W(CO)_6] \approx C_{MC,MC'}^{c}$ $[Mo(CO)_6]$ but the value for $Cr(CO)_6$ is considerably different.

The trans CO,C′O′ interaction $C_{CO,C'O'}^{t}$ is greater for solution than for vapor but the force constants $F_{CO,C'O'}^{t}$ show no statistical differences at 95% confidence; the relative standard deviations are much greater for $F_{CO,C'O'}^{t}$ than for $C_{CO,C'O'}^{t}$.

The rest of the stretching constants show no significant differences, either among the three molecules or between solution and vapor. It is hoped that many of the interaction constants can be used in solving the potential constant problem for other carbonyls. The average values are given in Table 5.2-5.

Thus, we find that the primary bending constants and most of the interaction constants are statistically the same for the three hexacarbonyls in gas and in solution. For some coordinates and interactions, the force constants are better determined, for others, the compliance constants are better determined. Because they are more unique, involving fewer assumptions, the compliance constants should be more transferable to metal carbonyls with similar structural moieties. However, at this time we do not have enough data to say which constants can be better transferred; it may be that for some coordinates, force constants are better while for others compliance constants are better.

Without further available evidence, if one wishes to estimate the potential constants of other metal carbonyls, it should be appropriate to transfer some of the smaller potential constants from the hexacarbonyls in order to make the problem tractable. Thus, the cis CO,C′O′ interaction, $F_{CO,C'O'}^{c}$ or $C_{CO,C'O'}^{c}$, is fairly well determined for the hexacarbonyls and without contrary evidence we would suppose that they could be transferred to other systems, such as the cis (90°) interactions of CO with C′O′ in $Fe(CO)_5$ and $Mn(CO)_5Br$. How suitable such generalities are remains to be seen. At present, the evidence is meager outside of the hexacarbonyls, but soon the normal coordinate problems should be at least partially

TABLE 5.2-5

STATISTICAL AVERAGES[a] AND STANDARD DEVIATIONS FOR F_{ij} AND C_{ij} OF $M(CO)_6$

Force constants			Compliance constants		
	F	σ		C	σ
F_{CO}	(17.015 to	17.310)[b]	C_{CO}	(0.06017 to	0.05897)[b]
F_{MC}	(1.906 to	2.385)	C_{MC}	(0.800 to	0.450)
F_{β}	0.46[c]	0.02[c]	C_{β}	2.49[c]	0.10[c]
$F_{\alpha} - F''_{\alpha\alpha}$	0.632	0.053	C_{α}	1.91	0.08
$F^{c}_{CO,C'O'}$	0.175	0.007	$C^{c}_{CO,C'O'}$	−0.00077	0.00002
$F^{t}_{CO,C'O'}$	0.039	0.016	$C^{t}_{CO,C'O'}$	(−0.0005 to	−0.0009)
$F^{c}_{MC,MC'}$	(−0.020 to	+0.052)	$C^{c}_{MC,MC'}$	(0.0023 to	−0.0069)
$F^{t}_{MC,MC'}$	(0.417 to	0.635)	$C^{t}_{MC,MC'}$	−0.114	0.004
$F_{MC,CO}$	0.758	0.016	$C_{MC,CO}$	−0.0238	0.0005
$F^{c}_{MC,C'O'}$	−0.088	0.006	$C^{c}_{MC,C'O'}$	0.00221	0.00012
$F^{t}_{MC,C'O'}$	−0.112	0.006	$C^{t}_{MC,C'O'}$	0.0097	0.0005
$F'_{\beta\beta}$	0.068	0.02	$C'_{\beta\beta}$	−0.25	0.10
$F''_{\beta\beta}$	0.023	0.005	$C''_{\beta\beta}$	−0.07	0.03
$F'''_{\beta\beta}$	−0.029	0.018	$C'''_{\beta\beta}$	0.24	0.06
$\bar{F}'_{\alpha\alpha} - F'''_{\alpha\alpha}$	0.140	0.026	$C''_{\alpha\alpha}$	−0.52	0.23
$F_{\alpha} - F''''_{\alpha\alpha}$	0.518	0.038	$C'_{\alpha\alpha}$	−0.22	0.03
$F'_{\alpha\beta}$	−0.110	0.013	$C'_{\alpha\beta}$	0.55	0.04
$F''_{\alpha\beta}$	−0.042	0.013	$C''_{\alpha\beta}$	0.09	0.04
$F'''_{\beta\beta}$	−0.015	0.008	$C'''_{\alpha\beta}$	0.03	0.04

[a] Units are the same as in Tables 5.2-2 and 5.2-4.

[b] The values in parentheses represent ranges of values for the six molecules. Statistically these constants are not all the same (at 95% confidence) for the three molecules in solution and vapour.

[c] The constants given with no parentheses are statistically the same for the three molecules in solution and vapor. The numbers are "best values" estimated, along with standard deviations (σ), as outlined by Hamilton (*6*).

solved for the well studied carbonyls $Fe(CO)_5$, $Mn(CO)_5Br$, etc. The force constants of nickel carbonyl have been determined (7) but its structure unfortunately has different bond angles than the octahedral structures. The interaction must be intimately connected with the bond angles. Nickel carbonyl is discussed later in this chapter.

5.3. INTERACTION COORDINATES OF THE HEXACARBONYLS

The interaction coordinates are readily calculated from the compliance constants; $(l)_k = C_{lk}/C_k$. As discussed previously, it was necessary to fix the constants $C_{CO,\alpha}$ and $C_{CO,\beta}$ in order to achieve a compliance constant solution for the F_{1u} symmetry block. First a solution was arrived at assuming $C_{CO,\alpha} = C_{CO,\beta} = 0$ as they are expected to be small. Then estimates of $(CO)_\alpha$ and $(CO)_\beta$ were made based on the assumption that

$$(CO)_\alpha = (CO)_{MC}(MC)_\alpha \qquad \text{and} \qquad (CO)_\beta = (CO)_{MC}(MC)_\beta \quad (5.3\text{-}1)$$

In the absence of more definite information, Eq. (5.3-1) is a reasonable approximation, as certainly to some extent a change in α_{ij} affects the

TABLE 5.3-1

INTERACTION COORDINATES FOR THE HEXACARBONYLS OF Cr, Mo, AND W

		μ^a	σ^a	
$(C_2O_2)_{C_1O_1}$	(cis)	−0.0129	0.0003	
$(C_6O_6)_{C_1O_1}$	(trans)	[b] −0.0137	0.0004	solution
		−0.0087	0.0008	gas
$(MC_1)_{C_1O_1}$		−0.401	0.009	
$(MC_2)_{C_1O_1}$	(cis)	0.037	0.002	
$(MC_6)_{C_1O_1}$	(trans)	0.164	0.009	
$(\beta_{21})_{C_1O_1}$		[−0.09][c] rad Å^{-1}		
$(\alpha_{12})_{C_1O_1}$		[−0.06][c] rad Å^{-1}		
$(MC_2)_{MC_1}$	(cis)	[b] 0.0041	0.0005	$Cr(CO)_6$
		−0.0128	0.0007	$Mo(CO)_6$, $W(CO)_6$
$(MC_6)_{MC_1}$	(trans)	−0.228	0.006	
$(C_1O_1)_{MC_1}$		−0.0446	0.0012	
$(C_2O_2)_{MC_1}$	(cis)	0.0040	0.0003	
$(C_6O_6)_{MC_1}$	(trans)	0.019	0.001	
$(\beta_{21})_{MC_1}$		0.21 rad Å^{-1}	0.04	
$(\alpha_{12})_{MC_1}$		0.16 rad Å^{-1}	0.03	
$(\alpha_{46})_{\alpha_{12}}$		0.27	0.05	
$(\alpha_{13})_{\alpha_{12}}$		−0.11	0.02	
$(\beta_{12})_{\alpha_{12}}$		0.30	0.05	
$(\beta_{62})_{\alpha_{12}}$		0.07	0.04	
$(\beta_{32})_{\alpha_{12}}$		0.03	0.03	
$(MC_1)_{\alpha_{12}}$		0.01 Å rad^{-1}	0.01	
$(C_1O_1)_{\alpha_{12}}$		[−0.002][c] Å rad^{-1}		
$(\beta_{62})_{\beta_{12}}$		−0.06	0.06	
$(\beta_{21})_{\beta_{12}}$		−0.03	0.01	
$(\beta_{32})_{\beta_{12}}$		0.12	0.04	
$(\alpha_{12})_{\beta_{12}}$		0.23	0.01	
$(\alpha_{26})_{\beta_{12}}$		0.05	0.03	
$(\alpha_{32})_{\beta_{12}}$		0.02	0.02	
$(MC_2)_{\beta_{12}}$		0.04 Å rad^{-1}	0.01	
$(C_2O_2)_{\beta_{12}}$		[−0.002][c] Å rad^{-1}		

[a] The statistical mean is given under μ; σ is the standard deviation.
[b] These values were statistically different, as indicated.
[c] These values depend on fixed compliance constants, as discussed in text.

MC_j bond which in turn induces an effect in the C_jO_j bond. In this manner the values -0.004 and $-0.005 \text{ rad}^2 \text{ Å}^{-1} \text{ mdyn}^{-1}$ were arrived at for $C_{\text{CO},\alpha}$ and $C_{\text{CO},\beta}$, respectively. With these values fixed, the other compliance constants were then recalculated. It was found that new estimates of $C_{\text{CO},\alpha}$ and $C_{\text{CO},\beta}$ were very little different from the previous values. The interaction coordinates as calculated from the compliance constants, averaged for the three hexacarbonyls, are listed in Table 5.3-1. An MCO angle change is designated β, while α designates a CMC angle change. The interaction coordinates are defined by Figure 5.2-1 and Table 5.2-3.

Interaction coordinates are intimately related to compliance constants [Eq. (1.10-7)] and should thus be equally transferable, if not more so. The main usefulness of interaction coordinates seems to be in relating the potential constants to electronic structure. Hopefully from a satisfactory molecular orbital theory one could calculate what coordinate displacements would minimize the energy if one particular coordinate is constrained to a displaced position. This would give a strong tie of vibrational theory to molecular orbital theory. At present such a correlation has not been made except in an extremely qualitative fashion. For example, if we stretch a CO bond, its antibonding pi orbitals become lower in energy and are thus more available for metal–CO pi-bonding,

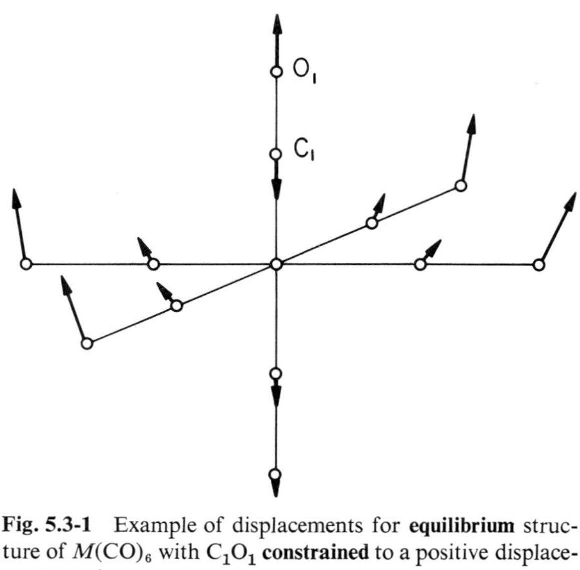

Fig. 5.3-1 Example of displacements for equilibrium structure of $M(CO)_6$ with C_1O_1 constrained to a positive displacement, as shown.

leading to a shortening of the adjacent metal carbon bond [thus, $(MC_1)_{C_1O_1} = -0.4$]. As an example of how the displacement of coordinates affect each other, in Fig. 5.3-1 we show, in an exaggerated fashion, the relative displacements of coordinates which minimizes the energy following a positive displacement of C_1O_1.

5.4. Nickel Carbonyl

Force constants for nickel carbonyl have been reported in the literature (*7*) as estimated from the vibrational spectra of the normal, the ^{13}C, and the ^{18}O species. The internal coordinate force constants are given in Table 5.4-1. The compliance constants yielding the same vibrational frequencies are given in Table 5.4-2. The various force constants are defined in (*8*).

5.5. A Comparison of the Potential Constants of the Hexacarbonyls with those of Nickel Carbonyl

The CO compliance constant is definitely smaller for $Ni(CO)_4$ than for $M(CO)_6$, indicating a stronger CO bond in $Ni(CO)_4$. This indicates less *M*–CO pi-bonding for the nickel compound than for the hexacarbonyls, because *M*–CO pi-bonding involves antibonding CO pi-orbitals, thus weakening the CO bond. The same conclusions follow from the F_{CO}. As evidenced by the values for F_{MC}, the Ni–C bond is about the same strength as the Mo–C and Cr–C bonds of the hexacarbonyls. The W–C bond is somewhat stronger. The conclusions regarding the bonding in these carbonyls are (*2*), then, that metal–CO pi-bonding is greater in the hexacarbonyls than in $Ni(CO)_4$, but that *M*–C sigma bonding is somewhat greater for $Ni(CO)_4$ than for $Mo(CO)_6$ or $Cr(CO)_6$. The W–C sigma bond is somewhat stronger than that of Mo–C or Cr–C. From thermochemical (*9*) and electron impact (*10*) measurements, the *M*–C dissociation energies are in the order W–C > Mo–C > Cr–C. Our measurements of both strengths are not in complete accord with the dissociation energies as we find W–C > Mo–C $\approx$ Cr–C.

The bending potential constants of $Ni(CO)_4$ are not well known; however, it is apparent that the *M*–C–O bending compliance constant is somewhat greater for $Ni(CO)_4$ than that for the hexacarbonyls meaning that Ni–C–O bends somewhat easier than *M*–C–O. This probably arises from the greater *M*–CO pi-bonding for $M(CO)_6$. The C*M*C bending compliance

TABLE 5.4-1

FORCE CONSTANTS OF NICKEL CARBONYL[a]

	Vapor		Solution
F_{CO}	17.85 ± 0.09		17.68 ± 0.09
F_{MC}	2.08 ± 0.10		2.09 ± 0.10
F_{β}	0.31 ± 0.08		0.31 ± 0.08
$F_{\alpha} - F_{\alpha\alpha'}$	0.19 (+0.03, −0.01)		0.25 (+0.04, −0.02)
$F_{\alpha} - F_{\alpha\alpha''}$	0.21 (+0.05, −0.02)		0.27 (+0.07, −0.03)
$F_{CO,C'O'}$	0.12 ± 0.04		0.13 ± 0.05
$F_{MC,MC'}$	0.10 ± 0.04		0.12 ± 0.03
$F_{MC,CO}$	0.52 ± 0.10		0.55 ± 0.08
$F_{MC,C'O'}$	−0.10 ± 0.04		−0.10 ± 0.04
$F_{\beta\beta'}$	−0.02 ± 0.06		−0.02 ± 0.06
$F_{\beta\beta''}$	0.08 ± 0.01		0.07 ± 0.01
$F_{\alpha\beta'}$		[0.06 ± 0.06]	
$F_{\alpha\beta''}$		[−0.01 ± 0.01]	
$F_{CO,\beta'}$		[0 ± 0.1]	
$F_{CO,\alpha'} - F_{CO,\alpha''}$		[0 ± 0.14]	
$F_{MC,\beta'}$	−0.06 (+0.02, −0.08)		−0.02 (+0.02, −0.02)
$F_{MC,\alpha'} - F_{MC,\alpha''}$		[0.14 ± 0.07]	

[a] From (*7*).

TABLE 5.4-2

COMPLIANCE CONSTANTS OF NICKEL CARBONYL

	Vapor	Solution
C_{CO}	0.05658	0.05716
C_{MC}	0.5475	0.5245
C_{β}		2.84
C_{α}		3.77
$C_{CO,C'O'}$	−0.00054	−0.00059
$C_{MC,MC'}$	−0.04109	−0.03802
$C_{MC,CO}$	−0.01678	−0.01699
$C_{MC,C'O'}$	0.00381	0.00385
$C'_{\beta\beta}$		−0.65
$C''_{\beta\beta}$		0.17
$C'_{\alpha\beta}$		−0.47
$C''''_{\alpha\beta}$		−0.14
$C'_{\alpha\alpha}$		−0.80
$C''_{\alpha\alpha}$		−0.56
$C'_{CO,\beta}$		−0.002
$C'_{CO,\alpha}$		0.006
$C'_{MC,\beta}$		−0.015
$C'_{MC,\alpha}$		−0.27

constant is much greater for $Ni(CO)_4$ than for $M(CO)_6$. Apparently the larger CMC angle makes it easier to bend the CMC bonds. It will be interesting to see how the 120° CMC angle of $Fe(CO)_5$ fits into the picture. We expect the compliance constant to be even greater than for CNiC.

The CO,C′O′ interactions are quite interesting. It was originally thought (*11, 12*) that $F^{t}_{CO,C'O'}$ should be about twice as great as $F^{c}_{CO,C'O'}$. This idea arose because it was felt that such interactions took place through the metal–CO pi-bond system; that is, a stretching of one CO bond would make its antibonding pi orbitals more available for bonding by the metal $d\pi$ electrons. This would lead to less pi-bonding in the other CO group. Such an effect would be twice as great for the trans CO group as for each cis CO group because the trans CO groups overlap the same two metal $d\pi$ orbitals while the cis CO groups share only one metal $d\pi$ orbital. This argument would apply also to compliance constants; that is, we might expect $C^{t}_{CO,C'O'}$ to be about twice as large as $C^{c}_{CO,C'O'}$. From Table 5.2-5 we see that $F^{t}_{CO,C'O'}$ is less than $F^{c}_{CO,C'O'}$ rather than twice as large. Also $C^{t}_{CO,C'O'}$ is equal to or less than $C^{c}_{CO,C'O'}$. Quite reasonably, the CO,C′O′ interaction of $Ni(CO)_4$ is somewhat less than the cis CO,C′O′ interaction of the hexacarbonyls. In fact, the CO,C′O′ interactions can be explained as dipole–dipole interactions of two CO groups (*13*). Thus a CO dipole interacts electrostatically with other CO dipoles in the molecule. When one CO bond is displaced, this dipole–dipole interaction will have an effect on the equilibrium configuration of the other CO bonds. It appears that the pi-bond system is not the major contribution to the CO–CO interactions.

The MC,MC' interaction of $Ni(CO)_4$ is definitely greater than the cis MC,MC' interaction of $M(CO)_6$ but considerably less than the trans MC,MC' interaction. This is to be expected; 109.5° interaction should be somewhat closer to 90° than to 180° interaction.

For the hexacarbonyls, the cis MC,C′O′ interaction force constant is somewhat less than the trans value though the difference is barely significant. The MC,C′O′ interaction for $Ni(CO)_4$ falls in between as expected. It appears that one could safely assign $F_{MC,C'O'}$ as -0.1 mdyn Å^{-1} for all MC,C′O′ interactions. The compliance constants show significant differences, indicating greater interaction for trans MC,C′O′ than for cis. Thus, $C^{cis}_{MC,C'O'} = 0.002$, $C^{tetra}_{MC,C'O'} = 0.004$, and $C^{trans}_{MC,C'O'} = 0.010$ Å mdyn^{-1}. The MC,CO interaction involving the same carbon atom is greater for the hexacarbonyls than for $Ni(CO)_4$, which is to be expected if it takes place primarily through the pi orbital system.

5.6. Simplified Force Fields for Metal Carbonyls

Cotton and Kraihanzel (*12*, *14*) introduced a simplified force field in which all force constants were neglected except the primary CO stretch and the CO,C′O′ interactions. Anharmonicity corrections were also neglected. For many carbonyls, such as the hexacarbonyls, it is then

TABLE 5.6-1

CO Force Constants and CO,C′O′ Interaction Constants for $M(CO)_6$ and $Ni(CO)_4$ in Solution[a] for Different Approximations

		General quadratic		CO factored field[b]	
		Harmonic	Anharmonic[c]	Harmonic	Anharmonic[c]
$Cr(CO)_6$	F_{CO}	17.04	16.52	16.95	16.45
	$F^{c}_{CO,C'O'}$	0.17	0.15	0.27	0.26
	$F^{t}_{CO,C'O'}$	0.08	0.28	0.35	0.54
$Mo(CO)_6$	F_{CO}	17.15	16.65	16.96	16.46
	$F^{c}_{CO,C'O'}$	0.17	0.16	0.29	0.27
	$F^{t}_{CO,C'O'}$	0.01	0.20	0.38	0.54
$W(CO)_6$	F_{CO}	17.02	16.52	16.84	16.35
	$F^{c}_{CO,C'O'}$	0.19	0.16	0.33	0.30
	$F^{t}_{CO,C'O'}$	0.11	0.27	0.41	0.57
$Ni(CO)_4$	F_{CO}	17.68	17.16	17.75	17.22
	$F_{CO,C'O'}$	0.13	0.18	0.29	0.34

[a] CCl_4 solution for all but $W(CO)_6$ for which CS_2 was used.
[b] Force constants calculated for $M(^{12}C^{16}O)_6$ assuming all other F_{ij} are zero.
[c] Force constants calculated in mdyn Å^{-1} using observed CO stretching frequencies, uncorrected for anharmonicity.

possible to calculate F_{CO} and the $F_{CO,C'O'}$ from the observed CO stretching frequencies alone. Such a procedure has been useful in making assignments of the various CO stretching vibrations of many metal carbonyls. However, the absolute values calculated using these approximations, as pointed out by Cotton (*15*), have little significance. A comparison of the true and approximate values is given in Table 5.6-1. It is apparent that use of the CO factored force field leads to values for $F^{c}_{CO,C'O'}$ and $F^{t}_{CO,C'O'}$ which are too large. Furthermore, use of observed anharmonic frequencies instead of harmonic frequencies leads to too large a value for $F^{t}_{CO,C'O'}$. The deviations in F_{CO} are not nearly so serious and one can hope to make reasonable comparisons among different molecules even with an approximate calculation. This arises because F_{CO} is calculated from a weighted average of the frequencies while the CO,C′O′ interactions are calculated from comparatively small differences.

Bor (*16*, *17*) has studied the infrared spectrum of some metal carbonyls partially enriched in ^{13}C. He has used these results to support force constant calculations. However, he has used a "CO factored" force field, without anharmonic corrections, so his results are no better than the fourth column of Table 5.6-1. The partially substituted species would be useful for potential constant calculation if one had an idea of the magnitude of the anharmonic corrections. However, one cannot transfer them from the normal species merely by an isotopic mass factor as change in symmetry may cause an unpredictable change in anharmonicity factors.

5.7. Iron Pentacarbonyl

The iron pentacarbonyl molecule $Fe(CO)_5$ has been the subject of several infrared investigations in attempts to determine whether the structure is a tetragonal pyramid (C_{4v} point group) or a trigonal bipyramid (D_{3h} point group). The results were not conclusive in this respect though they appeared to favor (*18*) the trigonal bipyramid. Hanson (*19*) and Donohue and Caron (*20*) have shown by X-ray studies that in the solid state $Fe(CO)_5$ is in the form of a slightly distorted trigonal bipyramid. The CO groups were shown (*20*) to be bonded to the iron through the carbon atom. Davis and Hanson (*21*) and Beagley et al. (*22*) have made electron diffraction studies of iron carbonyl and have come to the conclusion that the structure is a trigonal bipyramid. Furthermore, they find a significant difference in axial and equatorial Fe–C bond lengths, 1.806 ± 0.005 Å and 1.833 ± 0.004 Å respectively, according to Beagley et al. (*22*). Therefore we can expect, from Badger's rule (*23*), that the axial Fe–C force constant will be some 10–20% greater than the equatorial Fe–C force constant.

Edgell et al. (*18*) made a fairly complete assignment of the vibrational frequencies for $Fe(CO)_5$ by combining their infrared results with Raman studies of Stammreich et al. (*24*). Edgell et al. (*18*) give reference to the earlier work on $Fe(CO)_5$. The assignment (*18*) of the CO stretching frequencies were determined from vapor phase infrared spectra and liquid phase Raman spectra with no allowance for appreciable vapor-liquid frequency shift. Since that time Jones and McDowell (*25*) have revised the CO stretching frequency assignments after studies of the infrared spectrum of liquid $Fe(CO)_5$ in the CO stretching region. A good number of the other assignments are in doubt.

Recently McDowell et al. (*25*) have observed the spectra of $Fe(^{12}C^{16}O)_5$, $Fe(^{13}C^{16}O)_5$, and $Fe(^{12}C^{18}O)_5$. The most reasonable frequency assignment appears to be that in Table 5.7-1. Anharmonicity corrections have been determined for the CO stretching vibrations so the harmonic ω_{CO} are given also.

The calculation of potential constants has not been completed at this writing. A great difficulty arises because of the near equality of axial and equatorial CO stretching potential constants. However, by transferring many of the interaction constants of the hexacarbonyls to $Fe(CO)_5$, a satisfactory solution should be attainable.

TABLE 5.7-1

FUNDAMENTAL FREQUENCIES OF $Fe(CO)_5$ IN CS_2 SOLUTION

Representation and Designation		$Fe(^{12}C^{16}O)_5$	$Fe(^{13}C^{16}O)_5$	$Fe(^{12}C^{18}O)_5$
A_1'	ν_1, ω_1	2112.0, 2134.2	[2062.8, 2084.0][a]	[2066.3, 2087.5][a]
	ν_2, ω_2	2027.9, 2050.5	[1981.5, 2003.1][a]	[1983.8, 2005.4][a]
	ν_3	447	[440.2][b]	[429.6][b]
	ν_4	414.6	408.7	400.6
A_2'	ν_5			
A_2''	ν_6, ω_6	2018.8, 2049.2	1973.7, 2002.7	1972.5, 2001.5
	ν_7	614.6	603.8	611.3
	ν_8	475.3	465.8	466.1
	ν_9	~110	—	—
E'	ν_{10}, ω_{10}	1995.0, 2019.9	1949.9, 1973.7	1949.6, 1973.4
	ν_{11}	641.5	626.7	638.4
	ν_{12}	552.8	539.0	548.5
	ν_{13}	[426.6][c]	[416.5][c]	[422.8][c]
	ν_{14}	114.5	114.0	108.0
	ν_{15}	68		
E''	ν_{16}	492		
	ν_{17}	~374		
	ν_{18}	107		

[a] Estimated from combinations and anharmonic corrections.

[b] Estimated from combinations in gas phase and gas to solution shift.

[c] From CCl_4 solution and shift of ν_{13} for normal $Fe(CO)_5$ in going from CCl_4 to CS_2 solution.

5.8. Vibrational Spectra of Substituted Metal Carbonyls

A large number of substituted, or "mixed ligand," metal carbonyls have been prepared. Probably the most thoroughly studied of these are the manganese pentacarbonyl halides, $Mn(CO)_5X$ (*26–28*). However, complete vibrational analyses have not yet been made. Ottesen et al. (*30*) are investigating the infrared and Raman spectra of the three species

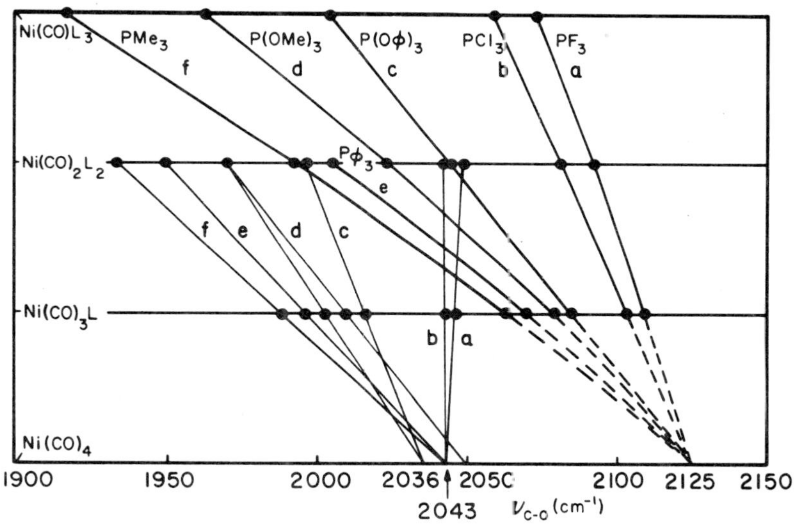

Fig. 5.8-1 Illustration of trend in CO frequencies for substituted nickel carbonyl. (Reprinted from (*31*) by courtsey of the bulletin de la Societé Chimique de France.)

$Mn(^{12}C^{16}O)_5Br$, $Mn(^{13}C^{16}O)_5Br$, and $Mn(^{12}C^{18}O)_5Br$. The results should yield a satisfactory frequency assignment and, with transferance from the hexacarbonyls, a unique set of potential constants.

It is possible to replace one or more of the CO groups in metal carbonyls by various neutral ligands such as PF_3, PR_3, $P(OR)_3$, $P\varphi_3$, etc. Bigorgne and coworkers have published a number of very interesting papers showing how the CO stretching frequency changes when such substitution is made. As an example, Fig. 5.8-1 is a reproduction from one of Bigorgne's papers (*31*), showing the change in CO stretching frequency when one or more CO groups of $Ni(CO)_4$ is replaced by $L = PF_3$, PCl_3, $P(O\varphi)_3$, $P(OMe)_3$, $P(\varphi)_3$, or $P(Me)_3$. The two CO stretching frequencies of $Ni(CO)_4$ occur at 2126 cm^{-1} (symmetric) and 2045 cm^{-1} (antisymmetric).

It is seen in Fig. 5.8-1 that as CO groups are replaced by L, both the symmetric and antisymmetric CO stretching frequencies decrease (except for the antisymmetric stretches with $L = PF_3$ and PCl_3). For $Ni(CO)L_3$ there is only one CO stretching frequency, and it is symmetric. There are several interesting observations from this data. The fact that for all but PCl_3 and PF_3 the antisymmetric and symmetric frequencies both decrease with increased substitution indicates that the primary force constant F_{CO} decreases also. Thus, for a group such as PMe_3 apparently metal–ligand pi backbonding is unfavorable so the pi bonding of the metal to the remaining CO groups is increased, resulting in lower F_{CO} values. On the other hand for PF_3 and PCl_3, the change in CO frequencies is relatively small indicating that they behave similar to CO groups with respect to metal–ligand backbonding. Thus PF_3 is a good acceptor of π electrons as pointed out by Wilkinson (*32*) many years ago. It is remarkable that the plots for the symmetric stretching vibrations extrapolate so well to the value for $Ni(CO)_4$.

A large number of other substituted carbonyls have been prepared by Bigorgne (*33–37*) and by many others. In general they observe the infrared active CO stretching frequencies of the products in solution because this is a fairly simple procedure and gives useful information about the symmetry of the molecule and the strengths of the CO bonds. Orgel (*38*) has discussed the assignments of infrared active CO stretching bands of substituted carbonyls from a consideration of the changing dipole moments of the individual CO groups and how they add vectorially for a given vibration. He shows that by combining intensity data with predictions of interaction constants one can generally arrive at a correct frequency assignment.

As discussed in Sect. 5.6, Cotton and Kraihanzel (*9*, *14*, *15*) went further than Orgel (*38*) and calculated "approximate" CO force constants for various substituted carbonyls. Their procedure was to neglect all force constants except the primary CO stretching force constants and the CO,CO interaction constants. Though it is not justified to interpret the "approximate" force constants, particularly the interaction constants, with respect to the nature of bonding, the approximate treatment has proven to be useful for assignment of vibrational frequencies. However, it appears that in most cases the correct assignments can be arrived at by the selection rules and intensity considerations discussed by Orgel (*38*) as well as qualitative predictions concerning the order of CO bond strengths in the molecule under study.

It seems now that before much more can be done with force constants of metal carbonyls the determination of anharmonicity correction and frequency shifts for fully enriched isotopic species must be pursued with vigor. Hopefully, some consistencies will be found which will make it possible to transfer anharmonicity constants, interaction constants, etc., from one carbonyl to another.

REFERENCES

1. W. Moffitt, *Proc. Roy. Soc.*, (London) **A196,** 524 (1949).
1a. J. Ladell, B. Post, and I. Fankuchen, *Acta Cryst.*, **5,** 795 (1952).
1b. J. Donohue and A. Caron, *Acta Cryst.*, **77,** 663 (1964).
2. L. H. Jones, R. S. McDowell, and M. Goldblatt, *Inorg. Chem.*, **8,** 2349 (1969).
3. J. M. Smith and L. H. Jones, *J. Mol. Spectry.*, **20,** 248 (1966).
4. R. G. Snyder and J. H. Schachtschneider, *Spectrochim. Acta*, **19,** 85 (1963).
5. W. C. Hamilton, *Statistics in Physical Science*, Ronald, New York, 1964.
6. *Ibid.*, p. 43.
7. L. H. Jones, R. S. McDowell, and M. Goldblatt, *J. Chem. Phys.*, **48,** 2663 (1968).
8. L. H. Jones, *J. Mol. Spectry.*, **5,** 133 (1960).
9. F. A. Cotton, A. K. Fischer, and G. Wilkinson, *J. Am. Chem. Soc.*, **81,** 800 (1959).
10. R. E. Winters and R. W. Kiser, *Inorg. Chem.*, **4,** 157 (1965).
11. L. H. Jones, *Advances in the Chemistry of Coordination Compounds* (S. Kirchner, ed.), Macmillan, New York, 1961, p. 398.
12. F. A. Cotton and C. S. Kraihanzel, *J. Am. Chem. Soc.*, **84,** 4432 (1962).
13. H. Haas and R. K. Sheline, *J. Chem. Phys.*, **47,** 2996 (1967).
14. C. S. Kraihanzel and F. A. Cotton, *Inorg. Chem.*, **2,** 533 (1963).
15. F. A. Cotton, *Inorg. Chem.*, **3,** 702 (1964).
16. G. Bor and G. Jung, *Inorg. Chim. Acta*, **3,** 69 (1969).
17. G. Bor, *Inorg. Chim. Acta*, **3,** 191 (1969).
18. W. F. Edgell, W. E. Wilson, and R. Summitt, *Spectrochim. Acta*, **19,** 863 (1963).
19. A. W. Hanson, *Acta Cryst.*, **15,** 930 (1962).
20. J. Donohue and A. Caron, *J. Phys. Chem.*, **70,** 603 (1966).
21. M. I. Davis and H. P. Hanson, *J. Phys. Chem.*, **69,** 3405 (1965).
22. B. Beagley, D. W. J. Cruickshank, P. M. Pinder, A. G. Robiette, and G. M. Sheldrick, *Acta Cryst.*, **B25,** 737 (1969).
23. R. M. Badger, *J. Chem. Phys.*, **3,** 710 (1935).
24. H. Stammreich, O. Sala, and Y. Tavares, *J. Chem. Phys.*, **30,** 856 (1959).
25. R. S. McDowell, L. H. Jones, and M. Goldblatt, unpublished work.
26. E. W. Abel and I. S. Butler, *J. Chem. Soc.*, 434 (1964).
27. H. D. Kaesz, R. Bau, D. Hendrickson, and J. M. Smith, *J. Am. Chem. Soc.*, **89,** 2844 (1967).
28. I. J. Hyams and E. R. Lippincott, *Spectrochim. Acta*, **25A,** 1845 (1969).
29. F. A. Cotton, A. Musco, and G. Yagupsky, *Inorg. Chem.*, **6,** 1357 (1967).
30. D. Ottesen, L. H. Jones, and M. Goldblatt, unpublished work.
31. M. Bigorgne, *Bull. Soc. Chim. France*, 1986 (1960).
32. G. Wilkinson, *J. Am. Chem. Soc.*, **73,** 5501 (1951).
33. M. Bigorgne, *Bull. Soc. Chim. France*, 295 (1963).
34. G. Bouquet and M. Bigorgne, *Bull. Soc. Chim. France*, 433 (1962).
35. M. Bigorgne, *Compt. Rend. Acad. Sci.*, **250,** 3484 (1960).
36. R. Poilblanc and M. Bigorgne, *Bull. Soc. Chim. France*, 1301 (1962).
37. M. Bigorgne, *J. Inorg. Nucl. Chem.*, **26,** 107 (1964).
38. L. Orgel, *Inorg. Chem.*, **1,** 25 (1962).

Appendix I

Determination of Anharmonicity Constants

For a polyatomic molecule with doubly degenerate (but not triply degenerate) vibrations, we can express the energy of a vibrational level as

$$G_{v_1,v_2,\ldots,\,l_1,l_1,\ldots} = \sum_i \omega_i[v_i + (d_i/2)]$$
$$+ \sum_i \sum_{j \geqslant i} [X_{ij}[v_i + (d_i/2)][v_j + (d_j/2)] + g_{ij} l_i l_j] \quad \text{(I-1)}$$

The integers, v_i and d_i, are the vibrational quantum number and degeneracy, respectively, of the frequencies, ω_i. The X_{ij} and g_{ij} are constants; g_{ij} is zero if either i or j represents a nondegenerate vibration. The l_i are internal angular momentum quantum numbers, equal to one of the values v_i, $v_i - 2, \ldots, 1$, or 0.

If we observe the frequency of a given vibrational absorption band, it tells us the energy difference between two vibrational energy levels. Thus, an observed frequency

$$[v_1\nu_1^{(l_1)} + v_2\nu_2^{(l_2)} + \cdots v_r\nu_r^{(l_r)}]_{\text{obs}}$$
$$= G_{v_1'v_2',\ldots;l_1',l_2',\ldots} - G_{v_1'',v_2'',\ldots;l_1'',l_2'',\ldots}$$
$$= \sum_i \omega_i(v_i' - v_i'') + \sum_i \sum_{j \geqslant i} [X_{ij}\{v_i'v_j' - v_i''v_j'' + (v_j' - v_j'')d_i/2$$
$$+ (v_i' - v_i'')d_j/2\} + g_{ij}\{l_i'l_j' - l_i''l_j''\}] \quad \text{(I-2)}$$

In this discussion, the value of l_i will be indicated by a superscript following ν_i in the observed frequency designation. The coefficient v_i of ν_i is the change in quantum number for the ith normal vibration which occurs during the indicated transition. Thus, for the nondegenerate fundamental frequency ν_1, we write

$$(\nu_1)_{\text{obs}} = G_{10\ldots}' - G_{00\ldots}' = \omega_1 + 2X_{11} + \sum_{j \neq 1} X_{1j}d_j/2$$

For the first overtone of ν_1,

$$(2\nu_1)_{\text{obs}} = G'_{20\ldots} - G''_{00\ldots} = 2\omega_1 + 6X_{11} + \sum_{j\neq i} X_{1j}d_j$$

These two relations lead to

$$X_{11} = [(2\nu_1)_{\text{obs}}/2] - (\nu_1)_{\text{obs}}$$

If enough vibrational transitions are observed, all the X_{ij}, ω_i, and g_{ij} can be determined in this manner. For example, if there are no triply degenerate vibrations, the following general solutions can be used.

$$X_{kk} = ([(2\nu_k)^{(2)}_{\text{obs}} + (2\nu_k)^{(0)}_{\text{obs}}]/4) - (\nu_k)_{\text{obs}}$$

$$g_{kk} = [(2\nu_k)^{(2)}_{\text{obs}} - (2\nu_k)^{(0)}_{\text{obs}}]/4$$

$$X_{kl} = [(\nu_k + 2\nu_l)^{(0)}_{\text{obs}} - (\nu_k)_{\text{obs}} - (2\nu_l)^{(0)}_{\text{obs}}]/2$$

$$g_{kl} = [(\nu_k + 2\nu_l^{(2)})_{\text{obs}} - (\nu_k + 2\nu_l^{(0)})_{\text{obs}} + (2\nu_l)^{(0)}_{\text{obs}} - (2\nu_l)^{(2)}_{\text{obs}}]/2$$

The superscripts give the values of l for degenerate vibrations. For nondegenerate vibrations, the superscript is to be ignored. Many other combinations can be made to yield the various constants; those above merely serve as an example.

For triply degenerate vibrations, the internal angular momentum term is more complex. It has been derived for tetrahedral molecules [see K. T. Hecht (*1*)].

A useful empirical relationship exists for calculating X_{ij} of isotopic molecules.

$$X_{ij}^{(k)} = X_{ij}^{(l)}\,\omega_i^{(k)}\,\omega_j^{(k)}/\omega_i^{(l)}\,\omega_j^{(l)}$$

In the few cases where sufficient data is available, this relationship is found to be quite satisfactory. It was proposed by L. G. Bonner (*2*) and B. T. Darling and D. M. Dennison (*3*). Such an expression does not apply for partially substituted isotopic species having different symmetries.

REFERENCES

1. K. T. Hecht, *J. Mol. Spectry.*, **5**, 355 (1960).
2. L. G. Bonner, *Phys. Rev.*, **46**, 458 (1934).
3. B. T. Darling and D. M. Dennison, *Phys. Rev.*, **57**, 128 (1940).

Appendix II

Normal Vibrations of Square Planar $M(\mathrm{CN})_4^{-x}$, D_{4h} Point Group

The character table for D_{4h} is given in Table IX-5, Appendix IX. To determine the number of vibrations in each irreducible representation we need the character for each symmetry operation, χ_R, as described in Sect. 1.4. The different types of internal coordinates as shown in Fig. II-1, are d_{AuC}, r_{CN}, $\beta_{M\mathrm{CN}}$, $\alpha_{\mathrm{C}M\mathrm{C}}$, $\pi_{M\mathrm{CN}}$, and $\epsilon_{\mathrm{C}M\mathrm{C}}$. The coordinates π and ϵ refer to out-of-plane bending. Table II-1 gives the characters for the various symmetry operations. Consider, for example, the in-plane bending of the MCN angles such as β_1 in Fig. II-1. There are four equivalent β angles (one at each carbon). The identity operation leaves them unchanged so $\chi_E^\beta = 4$. A rotation of $\pi/2$ or π about the four-fold axis changes all the β_i so $\chi_{C_4}^\beta = 0$ and $\chi_{C_2}^\beta = 0$. The C_2' operation is a rotation of π about a linear N–C–M–C–N axis. One of these will change β_1 to $-\beta_1$, β_3 to $-\beta_3$, β_4 to $-\beta_2$, and β_2 to $-\beta_4$; thus, $\chi_{C_2'}^\beta = -2$. In a similar fashion, the other χ_R^s are determined. The characters listed in Table II-1 are for each operation in the class designated by the column heading.

When these values are substituted in Eq. (1.4-1), we can determine the number of vibrations n_γ in each representation γ. For example, for $\gamma = A_{1g}$

$$n_{A_{1g}}^{\mathrm{CN}} = \tfrac{1}{16}[1 \times 4 \times 1 + 2 \times 2 \times 1 + 1 \times 4 \times 1 + 2 \times 2 \times 1]$$
$$= 1$$

$$n_{A_{1g}}^{\beta} = \tfrac{1}{16}[1 \times 4 \times 1 - 2 \times 2 \times 1 + 1 \times 4 \times 1 - 2 \times 2 \times 1]$$
$$= 0$$

Carrying on in this fashion, we arrive at the results of Table II-2. It is obvious that one cannot increase all CMC angles simultaneously as required to maintain A_{1g} symmetry. Therefore no real A_{1g} vibration can

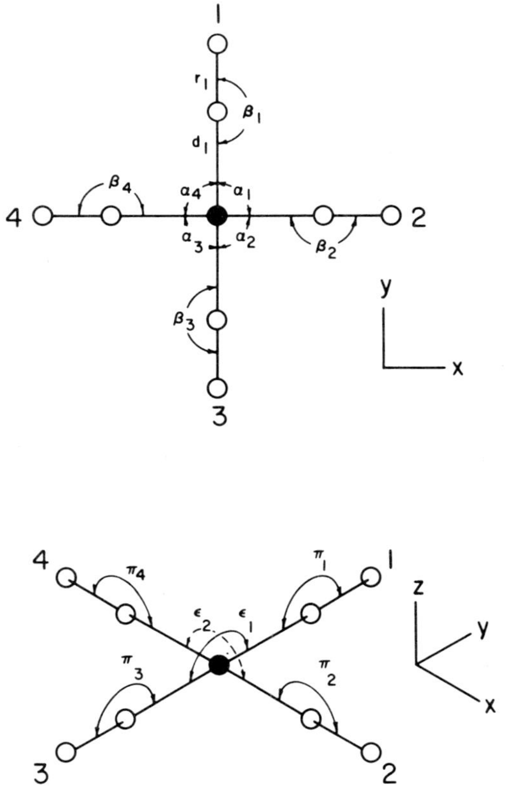

Fig. II-1 Internal coordinates of $M(CN)_4^{-x}$ of D_{4h} symmetry.

TABLE II-1

CHARACTERS FOR SYMMETRY OPERATIONS FOR $M(CN)_4$ OF D_{4h} POINT GROUP

χ_R^s	E	$2C_4$	C_2	$2C_2'$	$2C_2''$	i	$2S_4$	σ_h	$2\sigma_v$	$2\sigma_d$
d_{MC}	4	0	0	2	0	0	0	4	2	0
r_{CN}	4	0	0	2	0	0	0	4	2	0
β_{MCN}	4	0	0	−2	0	0	0	4	−2	0
α_{CMC}	4	0	0	0	2	0	0	4	0	2
π_{MCN}	4	0	0	−2	0	0	0	−4	2	0
ϵ_{CMC}	2	0	2	−2	0	−2	0	−2	2	0

involve bending of CMC angles. The α_{CMC} coordinate listed under A_{1g} is called redundant as it must equal zero. Therefore there are only two real A_{1g} vibrations. If we multiply the Σ in the last row by the degeneracy of that particular representation (A and B vibration have a degeneracy of one, while E vibrations have a degeneracy of two), the sum equals 21 which is $3N$-6 for the nine-atomic molecule. If we had not noticed the redundancy, we would have arrived at 22. Since this is one greater than $3N$-6, it tells us we must have a redundancy.

TABLE II-2

NUMBER OF VIBRATIONS FOR VARIOUS INTERNAL COORDINATES OF $M(CN)_4^{-x}$ CLASSIFIED BY REPRESENTATION

	A_{1g}	A_{2g}	B_{1g}	B_{2g}	E_g	A_{1u}	A_{2u}	B_{1u}	B_{2u}	E_u
r_{MC}	1	0	1	0	0	0	0	0	0	1
r_{CN}	1	0	1	0	0	0	0	0	0	1
β_{MCN}	0	1	0	1	0	0	0	0	0	1
α_{CMC}	1	0	0	1	0	0	0	0	0	1
π_{MCN}	0	0	0	0	1	0	1	0	1	0
ϵ_{CMC}	0	0	0	0	0	0	1	0	1	0
Σ	3 − 1[a] =2	1	2	2	1	0	2	0	2	4

[a] There is one A_{1g} redundancy which is a null vibration.

SYMMETRY COORDINATES

From Eq. (1.5-1), symmetry coordinates can be derived for $Au(CN)_4^-$.

$$S^{(\gamma)} = N \sum_R \chi_R^{(\gamma)} R(s_t) \tag{1.5-1}$$

First let us derive the symmetry coordinate $S_r^{A_{1g}}$. The transformation of r_1 under the various operations is given in Table II-3. Substitution of these values of $R(S_t)$ in Eq. (1.5-1), along with the characters $\chi_R^{(\gamma)}$ from Table IX-5, gives the symmetry coordinates S^{γ}

$$S_r^{A_{1g}} = N(2r_1 + 2r_2 + 2r_3 + 2r_4)$$

The symmetry coordinates should be normalized in the sense that for

$$S_t^{\gamma} = \sum_j C_{tj} S_j$$

$$\sum_j (C_{tj})^2 = 1$$

Thus, for $S_r^{A_{1g}}$

$$16N^2 = 1, \qquad N = \tfrac{1}{4}$$

$$S_r^{A_{1g}} = \tfrac{1}{2}[r_1 + r_2 + r_3 + r_4]$$

These r_i are displacement coordinates, signifying a small unit displacement of the CN bond lengths. Thus, an equal stretching of all four CN bonds maintains the molecular symmetry as required for the A_{1g} representation. This symmetry coordinate could, of course, be selected by inspection.

TABLE II-3

THE TRANSFORMATION COORDINATE $R(S_t)$ WHICH THE COORDINATES S_t GO TO UNDER OPERATION R

	E	$2C_4$	C_2	$2C_2'$	$2C_2''$	i	$2S_4$	σ_h	$2\sigma_v$	$2\sigma_d$
r_1	r_1	$r_2 + r_4$	r_3	$r_1 + r_3$	$r_2 + r_4$	r_3	$r_2 + r_4$	r_1	$r_1 + r_3$	$r_2 + r_4$
d_1	d_1	$d_2 + d_4$	d_3	$d_1 + d_3$	$d_2 + d_4$	d_3	$d_2 + d_4$	d_1	$d_1 + d_3$	$d_2 + d_4$
β_1	β_1	$\beta_2 + \beta_4$	β_3	$-\beta_1 - \beta_3$	$-\beta_2 - \beta_4$	β_3	$\beta_2 + \beta_4$	β_1	$-\beta_1 - \beta_3$	$-\beta_2 - \beta_4$
α_1	α_1	$\alpha_2 + \alpha_4$	α_3	$\alpha_2 + \alpha_4$	$\alpha_1 + \alpha_3$	α_3	$\alpha_2 + \alpha_4$	α_1	$\alpha_2 + \alpha_4$	$\alpha_1 + \alpha_3$
π_1	π_1	$\pi_2 + \pi_4$	π_3	$-\pi_1 - \pi_3$	$-\pi_2 - \pi_4$	$-\pi_3$	$-\pi_2 - \pi_4$	$-\pi_1$	$\pi_1 + \pi_3$	$\pi_2 + \pi_4$
ϵ_1	ϵ_1	$2\epsilon_2$	ϵ_1	$-2\epsilon_1$	$-2\epsilon_2$	$-\epsilon_1$	$-2\epsilon_2$	$-\epsilon_1$	$2\epsilon_1$	$2\epsilon_2$

As a less obvious case, let us derive the CN symmetry coordinate for the E_u representation. Using Table II-3 and the D_{4h} character table with Eq. (1.5-1), we find

$$S_r^{E_u} = N[2r_1 - 2r_3 - 2r_3 + 2r_1]$$
$$= \frac{1}{\sqrt{2}}[r_1 - r_3]$$

Thus one good E_u symmetry coordinate involves stretching of one CN bond with simultaneous compression of the opposite CN bond.

Since the E_u coordinates are degenerate, for the other E_u internal coordinate sets we must use linear combinations which have the same symmetry as r_1, which was operated on to derive $S_r^{E_u}$. For β such a combination would be $\beta_4 - \beta_2$. Following through with Eq. (1.5-1) and transformations for β_4 and β_2 analogous to those for β_1 in Table II-3, we arrive at

$$S_\beta^{E_u} = \frac{1}{\sqrt{2}}(\beta_4 - \beta_2)$$

Proceeding in this fashion, we arrive at the 17 symmetry coordinates of Table II-4.

TABLE II-4

SYMMETRY COORDINATES OF SQUARE PLANAR $M(CN)_4^{-x}$

$$S_r^{A_{1g}} = \tfrac{1}{2}(r_1 + r_2 + r_3 + r_4)$$
$$S_d^{A_{1g}} = \tfrac{1}{2}(d_1 + d_2 + d_3 + d_4)$$
$$S_\alpha^{A_{2g}} = \tfrac{1}{2}(\alpha_1 + \alpha_2 + \alpha_3 + \alpha_4)$$
$$S_\beta^{A_{2g}} = \tfrac{1}{2}(\beta_1 + \beta_2 + \beta_3 + \beta_4)$$
$$S_r^{B_{1g}} = \tfrac{1}{2}(r_1 - r_2 + r_3 - r_4)$$
$$S_d^{B_{1g}} = \tfrac{1}{2}(d_1 - d_2 + d_3 - d_4)$$
$$S_\beta^{B_{2g}} = \tfrac{1}{2}(\beta_1 - \beta_2 + \beta_3 - \beta_4)$$
$$S_\alpha^{B_{2g}} = \tfrac{1}{2}(\alpha_1 - \alpha_2 + \alpha_3 - \alpha_4)$$
$$S_\pi^{E_g} = \frac{1}{\sqrt{2}}(\pi_1 - \pi_3)$$
$$S_\pi^{A_{2u}} = \tfrac{1}{2}(\pi_1 + \pi_2 + \pi_3 + \pi_4)$$
$$S_\epsilon^{A_{2u}} = \frac{1}{\sqrt{2}}(\epsilon_1 + \epsilon_2)$$
$$S_\pi^{B_{2u}} = \tfrac{1}{2}(\pi_1 - \pi_2 + \pi_3 - \pi_4)$$
$$S_\epsilon^{B_{2u}} = \frac{1}{\sqrt{2}}(\epsilon_1 - \epsilon_2)$$
$$S_r^{E_u} = \frac{1}{\sqrt{2}}(r_1 - r_3)$$
$$S_d^{E_u} = \frac{1}{\sqrt{2}}(d_1 - d_3)$$
$$S_\beta^{E_u} = \frac{1}{\sqrt{2}}(\beta_4 - \beta_2)$$
$$S_\alpha^{E_u} = \tfrac{1}{2}(\alpha_1 - \alpha_2 - \alpha_3 + \alpha_4)$$

G MATRIX FOR $M(CN)_4^{-x}$ OF D_{4h} SYMMETRY

As in Eq. (1.6-2), we wish to express the symmetry coordinates in terms of Cartesian displacement coordinates. For example, consider the two A_{1g} stretching coordinates ($S_\alpha^{A_{1g}}$ is equal to zero as mentioned above).

$$S_r^{A_{1g}} = \tfrac{1}{2}[Y_{N_1} - Y_{C_1} + X_{N_2} - X_{C_2} - Y_{N_3} + Y_{C_3} - X_{N_4} + X_{C_4}]$$
$$S_d^{A_{1g}} = \tfrac{1}{2}[Y_{C_1} - Y_M + X_{C_2} - X_M - Y_{C_3} - Y_M - X_{C_4} + X_M]$$

Thus,

$$B_{rY_{N_1}}^{A_{1g}} = \tfrac{1}{2}, \qquad B_{rY_{C_1}}^{A_{1g}} = -\tfrac{1}{2}, \qquad B_{rY_{N_2}}^{A_{1g}} = 0, \qquad \text{etc.}$$

Each of the Cartesian coordinates are small unit displacements. Thus, Y_{N_1} stands for ΔY_{N_1}.

We can then substitute the various B elements in Eq. (1.6-8). We find

$$\begin{aligned}
G_{rr}^{A_{1g}} &= 1/m_{\mathrm{N}}[(\tfrac{1}{2})(\tfrac{1}{2}) + (\tfrac{1}{2})(\tfrac{1}{2}) + (-\tfrac{1}{2})(-\tfrac{1}{2}) + (-\tfrac{1}{2})(-\tfrac{1}{2})] \\
&\quad + 1/m_{\mathrm{C}}[(-\tfrac{1}{2})(-\tfrac{1}{2}) + (-\tfrac{1}{2})(-\tfrac{1}{2}) + (\tfrac{1}{2})(\tfrac{1}{2}) + (\tfrac{1}{2})(\tfrac{1}{2})] \\
&= \mu_{\mathrm{N}} + \mu_{\mathrm{C}} \\
\mu_i &= 1/m_i \\
G_{rd}^{A_{1g}} &= 1/m_{\mathrm{N}}[(\tfrac{1}{2})(0) + (\tfrac{1}{2})(0) + (-\tfrac{1}{2})(0) + (-\tfrac{1}{2})(0)] \\
&\quad + 1/m_{\mathrm{C}}[(-\tfrac{1}{2})(\tfrac{1}{2}) + (-\tfrac{1}{2})(\tfrac{1}{2}) + (\tfrac{1}{2})(-\tfrac{1}{2}) + (\tfrac{1}{2})(-\tfrac{1}{2})] \\
&= -\mu_c
\end{aligned}$$

Similarly

$$G_{dd}^{A_{1g}} = \mu_c$$

Slightly more complicated are the bending coordinates. Consider the B_{2g} block.

$$\begin{aligned}
S_{\beta}^{B_{2g}} &= \tfrac{1}{2}(\beta_1 - \beta_2 + \beta_3 - \beta_4) \\
S_{\alpha}^{B_{2g}} &= \tfrac{1}{2}(\alpha_1 - \alpha_2 + \alpha_3 - \alpha_4) \\
S_{\beta}^{B_{2g}} &= \frac{1}{2}\Bigg[-\frac{X_{\mathrm{N}_1}}{r} + \frac{X_{\mathrm{C}_1}}{r} + \frac{X_{\mathrm{C}_1}}{d} - \frac{X_M}{d} - \frac{Y_{\mathrm{N}_2}}{r} + \frac{Y_{\mathrm{C}_2}}{r} \\
&\quad + \frac{Y_{\mathrm{C}_2}}{d} - \frac{Y_M}{d} + \frac{X_{\mathrm{N}_3}}{r} - \frac{X_{\mathrm{C}_3}}{r} - \frac{X_{\mathrm{C}_3}}{d} + \frac{X_M}{d} \\
&\quad + \frac{Y_{\mathrm{N}_4}}{r} - \frac{Y_{\mathrm{C}_4}}{r} - \frac{Y_{\mathrm{C}_4}}{d} + \frac{Y_M}{d}\Bigg] \\
S_{\alpha}^{B_{2g}} &= \frac{1}{2}\Bigg[-\frac{X_{\mathrm{C}_1}}{d} - \frac{Y_{\mathrm{C}_2}}{d} + \frac{X_M}{d} + \frac{Y_M}{d} - \frac{Y_{\mathrm{C}_2}}{d} + \frac{X_{\mathrm{C}_3}}{d} + \frac{Y_M}{d} \\
&\quad - \frac{X_M}{d} + \frac{X_{\mathrm{C}_3}}{d} + \frac{Y_{\mathrm{C}_4}}{d} - \frac{X_M}{d} - \frac{Y_M}{d} - \frac{X_{\mathrm{C}_1}}{d} \\
&\quad + \frac{Y_{\mathrm{C}_4}}{d} + \frac{X_M}{d} - \frac{Y_M}{d}\Bigg]
\end{aligned}$$

We may now write the B matrix:

$B^{B_{2g}}$	$S_\beta^{B_{2g}}$	$S_\alpha^{B_{2g}}$
X_{N_1}	$-\frac{1}{2r}$	0
Y_{N_2}	$-\frac{1}{2r}$	0
X_{N_3}	$+\frac{1}{2r}$	0
Y_{N_4}	$+\frac{1}{2r}$	0
X_{C_1}	$+\frac{1}{2}\left(\frac{1}{r}+\frac{1}{d}\right)$	$-\frac{1}{d}$
Y_{C_2}	$+\frac{1}{2}\left(\frac{1}{r}+\frac{1}{d}\right)$	$-\frac{1}{d}$
X_{C_3}	$-\frac{1}{2}\left(\frac{1}{r}+\frac{1}{d}\right)$	$+\frac{1}{d}$
Y_{C_4}	$-\frac{1}{2}\left(\frac{1}{r}+\frac{1}{d}\right)$	$+\frac{1}{d}$
all other ξ_i	0	0

Then, using Eq. (1.6-3), we find

$$G_{\beta\beta}^{B_{2g}} = \mu_N \frac{1}{r^2} + \mu_C\left(\frac{1}{r}+\frac{1}{d}\right)^2$$

$$G_{\alpha\alpha}^{B_{2g}} = 4\mu_C\left(\frac{1}{d}\right)^2$$

$$G_{\beta\alpha}^{B_{2g}} = -2\mu_C \frac{1}{d}\left(\frac{1}{r}+\frac{1}{d}\right)$$

Continuing in this manner we arrive at the G matrix for square planar $M(CN)_4^{-x}$.

TABLE II-5

G MATRIX FOR $M(\mathrm{CN})_4^{-x}$ OF D_{4h} SYMMETRY

$$G^{A_{1g}}_{rr} = \mu_N + \mu_C$$
$$G^{A_{1g}}_{dd} = \mu_C$$
$$G^{A_{1g}}_{rd} = -\mu_C$$
$$G^{A_{2g}}_{\beta\beta} = \mu_N r^{-2} + \mu_C(r^{-1} + d^{-1})^2$$

$$G^{B_{1g}}_{rr} = \mu_N + \mu_C$$
$$G^{B_{1g}}_{dd} = \mu_C$$
$$G^{B_{1g}}_{rd} = -\mu_C$$

$$G^{B_{2g}}_{\beta\beta} = \mu_N r^{-2} + \mu_C(r^{-1} + d^{-1})^2$$
$$G^{B_{2g}}_{\alpha\alpha} = 4\mu_C d^{-2}$$
$$G^{B_{2g}}_{\beta\alpha} = -2\mu_C d^{-1}(r^{-1} + d^{-1})$$

$$G^{E_g}_{\pi\pi} = \mu_N r^{-2} + \mu_C(r^{-1} + d^{-1})^2$$
$$G^{A_{2u}}_{\pi\pi} = \mu_N r^{-2} + \mu_C(r^{-1} + d^{-1})^2 + 4\mu_M d^{-2}$$
$$G^{A_{2u}}_{\epsilon\epsilon} = 2\mu_C d^{-2} + 8\mu_M d^{-2}$$
$$G^{A_{2u}}_{\pi\epsilon} = -\sqrt{2}[\mu_C d^{-1}(r^{-1} + d^{-1}) + 4\mu_M d^{-2}]$$

$$G^{B_{2u}}_{\pi\pi} = \mu_N r^{-2} + \mu_C(r^{-1} + d^{-1})^2$$
$$G^{B_{2u}}_{\epsilon\epsilon} = 2\mu_C d^{-2}$$
$$G^{B_{2u}}_{\pi\epsilon} = \sqrt{2}\,\mu_C d^{-1}(r^{-1} + d^{-1})$$

$$G^{E_u}_{rr} = \mu_N + \mu_C$$
$$G^{E_u}_{dd} = \mu_C + 2\mu_M$$
$$G^{E_u}_{\beta\beta} = \mu_N r^{-2} + \mu_C(r^{-1} + d^{-1})^2 + 2\mu_M d^{-2}$$
$$G^{E_u}_{\alpha\alpha} = 2\mu_C d^{-2} + \mu_M d^{-2}$$
$$G^{E_u}_{rd} = -\mu_C$$
$$G^{E_u}_{r\beta} = 0$$
$$G^{E_u}_{r\alpha} = 0$$
$$G^{E_u}_{d\beta} = -2\mu_M d^{-1}$$
$$G^{E_u}_{d\alpha} = -2\sqrt{2}\,\mu_M d^{-1}$$
$$G^{E_u}_{\beta\alpha} = \sqrt{2}[\mu_C d^{-1}(r^{-1} + d^{-1}) + 2\mu_M d^{-2}]$$

μ_N = reciprocal of mass of N, etc.
r = CN distance $\quad d$ = *M*C distance

F MATRIX

To determine the symmetry force constants in terms of internal coordinate force constants, we use Eq. (1.6-11) and (1.6-12). For example,

$$S_r^{B_{1g}} = \tfrac{1}{2}(r_1 - r_2 + r_3 - r_4)$$

from Eq. (1.6-11).

$$F_r^{B_{1g}} = \frac{\tfrac{1}{2}(F_r - F_{r_1r_2} + F_{r_1r_3} - F_{r_1r_4})}{\tfrac{1}{2}}$$
$$= F_r - 2F_{rr}^{\mathrm{c}} + F_{rr}^{\mathrm{t}}$$

In this relation F_{rr}^{c} is for *r–r* interaction at 90° (cis) while F_{rr}^{t} is for opposite *r–r* interaction (trans). For an off-diagonal constant, we use Eq. (1.6-12).

TABLE II-6

SYMMETRY FORCE CONSTANTS OF $M(CN)_4^{-x}$ (D_{4h})

$F_r^{A_{1g}} = F_r + 2F_{rr}^c + F_{rr}^t$	$F_\pi^{B_{2u}} = F_\pi - 2F_{\pi\pi}^c + F_{\pi\pi}^t$
$F_d^{A_{1g}} = F_d + 2F_{dd}^c + F_{dd}^t$	$F_\epsilon^{B_{2u}} = F_\epsilon - F'_{\epsilon\epsilon}$
$F_{rd}^{A_{1g}} = F_{rd} + 2F_{rd}^c + F_{rd}^t$	$F_{\pi\epsilon}^{B_{2u}} = \sqrt{2}(F_{\pi\epsilon} - F'_{\pi\epsilon})$
$F_\beta^{A_{2g}} = F_\beta + 2F'_{\beta\beta} + F''_{\beta\beta}{}^a$	$F_r^{E_u} = F_r - F_{rr}^t$
$F_r^{B_{1g}} = F_r - 2F_{rr}^c + F_{rr}^t$	$F_d^{E_u} = F_d - F_{dd}^t$
$F_d^{B_{1g}} = F_d - 2F_{dd}^c + F_{dd}^t$	$F_\beta^{E_u} = F_\beta - F''_{\beta\beta}$
$F_{rd}^{B_{1g}} = F_{rd} - 2F_{rd}^c + F_{rd}^t$	$F_\alpha^{E_u} = F_\alpha - F''_{\alpha\alpha}$
$F_\beta^{B_{2g}} = F_\beta - 2F'_{\beta\beta} + F''_{\beta\beta}$	$E_{rd}^{E_u} = F_{rd} - F_{rd}^t$
$F_\alpha^{B_{2g}} = F_\alpha - 2F'_{\alpha\alpha} + F''_{\alpha\alpha}$	$F_{r\beta}^{E_u} = 2F_{r\beta}$
$F_{\beta\alpha}^{B_{2g}} = 2(F_{\beta\alpha} + F'_{\beta\alpha})$	$F_{r\alpha}^{E_u} = \sqrt{2}(F_{r\alpha} - F'_{r\alpha})$
$F_\pi^{E_g} = F_\pi - F''_{\pi\pi}$	$F_{d\beta}^{E_u} = 2F_{d\beta}$
	$F_{d\alpha}^{E_u} = \sqrt{2}(F_{d\alpha} - F'_{d\alpha})$
$F_\pi^{A_{2u}} = F_\pi + 2F'_{\pi\pi} + F''_{\pi\pi}$	$F_{\beta\alpha}^{E_u} = \sqrt{2}(F_{\beta\alpha} - F'_{\beta\alpha})$
$F_\epsilon^{A_{2u}} = F_\epsilon + F'_{\epsilon\epsilon}$	
$F_{\pi\epsilon}^{A_{2u}} = \sqrt{2}[F_{\pi\epsilon} + F'_{\pi\epsilon}]$	

[a] The single primes and double primes refer to interaction at 90° and 180°, respectively, except for the following:

$F_{\beta\alpha}$	α_1 with β_1
$F'_{\beta\alpha}$	α_1 with β_3
$F_{\pi\epsilon}$	π_1 with ϵ_1
$F'_{\pi\epsilon}$	π_1 with ϵ_2
$F_{r\beta}$, $F_{d\beta}$	r_1, d_1 with β_4
$F'_{r\alpha}$, $F'_{d\alpha}$	r_1, d_1 with α_1
$F'_{\alpha\alpha}$	α_1 with α_2
$F''_{\alpha\alpha}$	α_1 with α_3

For example,

$$F_{\pi\epsilon}^{A_{2u}} = \frac{\frac{1}{2}[F_{\pi_1\epsilon_1} + F_{\pi_2\epsilon_1} + F_{\pi_3\epsilon_1} + F_{\pi_4\epsilon_1}]}{1/\sqrt{2}}$$

$$= \frac{1}{\sqrt{2}}[2F_{\pi\epsilon} + 2F'_{\pi\epsilon}] = \sqrt{2}(F_{\pi\epsilon} + F'_{\pi\epsilon})$$

$F_{\pi\epsilon}$ is for interaction of π with ϵ in the same plane. $F'_{\pi\epsilon}$ is for interaction of π with ϵ in a perpendicular plane. Applying Eqs. (1.6-11) and (1.6-12) in this fashion leads to the force constants of Table II-6.

Appendix III

Normal Coordinate Treatment of Linear *ABC* Using Cartesian Coordinates

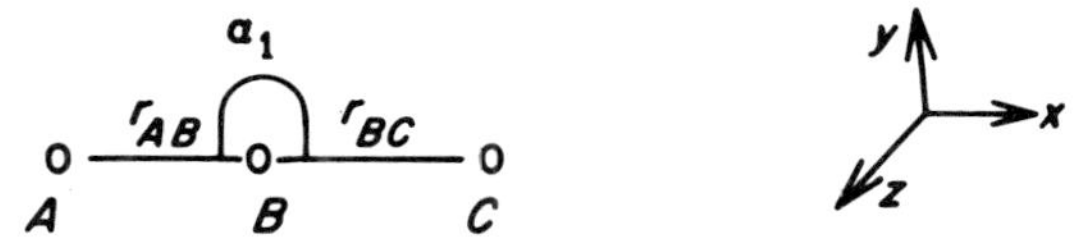

$$2V = F_{AB}(\Delta r_{AB})^2 + F_{BC}(\Delta r_{BC})^2 + 2F_{AB,BC}(\Delta r_{AB})(\Delta r_{BC}) + F_\alpha[(\Delta\alpha_1)^2 + (\Delta\alpha_2)^2]$$

$\Delta\alpha_1$ is a change in *ABC* angle in the *xy* plane; $\Delta\alpha_2$ is a change in *ABC* angle in the *xz* plane. We do not have to include interaction terms like $F_{AB,\alpha}$ and $F_{\alpha,\alpha}$. This is obvious because in the quadratic approximation stretching *AB* cannot cause a displacement of α_1 or α_2. The symmetry requires that. Similarly, a displacement of α_1 cannot cause a displacement of α_2.

$$2V = F_{AB}(\Delta x_B - \Delta x_A)^2 + F_{BC}(\Delta x_C - \Delta x_B)^2 + 2F_{AB,BC}(\Delta x_B - \Delta x_A)(\Delta x_C - \Delta x_B) + F_\alpha\left[\left(\frac{\Delta y_B - \Delta y_C}{r_{BC}} + \frac{\Delta y_B - \Delta y_A}{r_{AB}}\right)^2 + \left(\frac{\Delta z_B - \Delta z_C}{r_{BC}} + \frac{\Delta z_B - \Delta z_A}{r_{AB}}\right)^2\right] \quad \text{(III-1)}$$

$$2T = M_A(\dot{x}_A^2 + \dot{y}_A^2 + \dot{z}_A^2) + M_B(\dot{x}_B^2 + \dot{y}_B^2 + \dot{z}_B^2) + M_C(\dot{x}_C^2 + \dot{y}_C^2 + \dot{z}_C^2)$$

Lagrange's Equations of Motion are

$$\frac{d}{dt}\left(\frac{\partial T}{\partial \dot{q}_i}\right) + \frac{\partial V}{\partial q_i} = 0 \tag{III-2}$$

q_i is a particular Cartesian displacement coordinate, Δx_A, Δy_A, $\Delta x_B, \ldots$; $\dot{q}_i = dq_i/dt$. Application of (III-2) to (III-1) for each of the nine Cartesian displacement coordinates leads to a series of nine equations as

$$M_A\ddot{x}_A - F_{AB}(\Delta x_B - \Delta x_A) - F_{AB,BC}(\Delta x_C - \Delta x_B) = 0$$

$$M_B\ddot{x}_B + F_{AB}(\Delta x_B - \Delta x_A) - F_{BC}(\Delta x_C - \Delta x_B)$$
$$+ F_{AB,BC}(\Delta x_C - 2\Delta x_B \Delta x_A) = 0$$

$$M_C\ddot{x}_C + F_{BC}(\Delta x_C - \Delta x_B) + F_{AB,BC}(\Delta x_B - \Delta x_A) = 0$$

$$M_A\ddot{y}_A - \frac{F_\alpha}{r_{AB}}\left(\frac{\Delta y_B - \Delta y_C}{r_{BC}} + \frac{\Delta y_B - \Delta y_A}{r_{AB}}\right) = 0$$

$$M_B\ddot{y}_B + F_\alpha\left(\frac{1}{r_{AB}} + \frac{1}{r_{BC}}\right)\left(\frac{\Delta y_B - \Delta y_C}{r_{BC}} + \frac{\Delta y_B - \Delta y_A}{r_{AB}}\right) = 0 \tag{III-3}$$

$$M_C\ddot{y}_C - \frac{F_\alpha}{r_{BC}}\left(\frac{\Delta y_B - \Delta y_C}{r_{BC}} + \frac{\Delta y_B - \Delta y_A}{r_{AB}}\right) = 0$$

$$M_A\ddot{z}_A - \frac{F_\alpha}{r_{AB}}\left(\frac{\Delta z_B - \Delta z_C}{r_{BC}} + \frac{\Delta z_B - \Delta z_A}{r_{AB}}\right) = 0$$

$$M_B\ddot{z}_B + F_\alpha\left(\frac{1}{r_{BC}} + \frac{1}{r_{AB}}\right)\left(\frac{\Delta z_B - \Delta z_C}{r_{BC}} + \frac{\Delta z_B - \Delta z_A}{r_{AB}}\right) = 0$$

$$M_C\ddot{z}_C - \frac{F_\alpha}{r_{BC}}\left(\frac{\Delta z_B - \Delta z_C}{r_{BC}} + \frac{\Delta z_B - \Delta z_A}{r_{AB}}\right) = 0$$

$$\ddot{x}_i = d^2(\Delta x_i)/dt^2$$

Thus we have a series of nine differential equations in nine unknowns. A solution is

$$x_i = e^{2\pi i\nu t}$$

$$\ddot{x}_i = -4\pi^2\nu^2 e^{2\pi i\nu t} = -\lambda x_i$$

$$\lambda = 4\pi^2\nu^2$$

This solution is periodic, with frequency ν. It is the harmonic oscillator solution as expected from the quadratic potential function of (III-1).

Thus, for each $\ddot{x}_i$ in (III-3) we substitute $-\lambda\,\Delta x_i$. The set of equations of motion then becomes

$$
\begin{aligned}
&(F_{AB} - M_A\lambda)\,\Delta x_A - (F_{AB} - F_{AB,BC})\,\Delta x_B - F_{AB,BC}\,\Delta x_C = 0\\
&-(F_{AB} - F_{AB,BC})\,\Delta x_A + (F_{AB} + F_{BC} - 2F_{AB,BC} - M_B\lambda)\,\Delta x_B\\
&\qquad\qquad - (F_{BC} - F_{AB,BC})\,\Delta x_C = 0\\
&-F_{AB,BC}\,\Delta x_A - (F_{BC} - F_{AB,BC})\,\Delta x_B + (F_{BC} - M_C\lambda)\,\Delta x_C = 0\\
&\left(\frac{F_\alpha}{r_{AB}^2} - M_A\lambda\right)\Delta y_A - \frac{F_\alpha}{r_{AB}}\left[\frac{1}{r_{BC}} + \frac{1}{r_{AB}}\right]\Delta y_B + \frac{F_\alpha}{r_{AB}r_{BC}}\Delta y_C = 0\\
&-\frac{F_\alpha}{r_{AB}}\left(\frac{1}{r_{AB}} + \frac{1}{r_{BC}}\right)\Delta y_A + \left[F_\alpha\left(\frac{1}{r_{AB}} + \frac{1}{r_{BC}}\right)^2 - M_B\lambda\right]\Delta y_B\\
&\qquad\qquad - \frac{F_\alpha}{r_{BC}}\left(\frac{1}{r_{AB}} + \frac{1}{r_{BC}}\right)\Delta y_C = 0\\
&\frac{F_\alpha}{r_{BC}r_{AB}}\Delta y_A - \frac{F_\alpha}{r_{BC}}\left(\frac{1}{r_{AB}} + \frac{1}{r_{BC}}\right)\Delta y_B + \left(\frac{F_\alpha}{r_{BC}^2} - M_C\lambda\right)\Delta y_C = 0
\end{aligned}
\tag{III-4}
$$

and equations for Δz_i analogous to those for Δy_i. Only special values of λ give nonvanishing solutions to (III-4). These special values of λ are those which satisfy the *secular determinant* made up of the coefficients of the Δx_i.

$$
\begin{vmatrix}
F_{AB}-M_A\lambda & -(F_{AB}-F_{AB,BC}) & -F_{AB,BC} & 0\ldots\\
-(F_{AB}-F_{AB,BC}) & (F_{AB}+F_{BC}-2F_{AB,BC}-M_B\lambda) & -(F_{BC}-F_{AB,BC}) & 0\ldots\\
-F_{AB,BC} & -(F_{BC}-F_{AB,BC}) & (F_{BC}-M_C\lambda) & 0\ldots\\
0 & 0 & 0 & \frac{F_\alpha}{r_{AB}{}^2}-M_A\lambda\ldots
\end{vmatrix} = 0
\tag{III-5}
$$

We thus have a 9×9 determinant which, in this case, however, breaks up into three 3×3 blocks. Therefore each 3×3 can be treated independently. This is actually obvious from Eq. (III-4). The terms in Δx_A, Δx_B, and Δx_C appear only in the first three equations. Therefore, to solve for the motions Δx_A, Δx_B, and Δx_C, we need only the first three simultaneous equations. Similarly, only the second three are required to solve for Δy_A, Δy_B, and Δy_C.

Therefore we first look at the determinant of the coefficients in the first three equations, given by the first three rows and columns of (III-5). By expanding this 3×3 determinant, equating to zero, and solving for λ, we arrive at a relation between force constants and λ (from which we can calculate the vibrational frequencies ν). The values of ν which satisfy this *secular equation* are the vibrational frequencies.

Expansion of the 3×3 block in the upper left-hand corner of (III-5) leads to the relation

$$\lambda^3 - \left[F_{AB}\left(\frac{1}{M_A} + \frac{1}{M_B}\right) + F_{BC}\left(\frac{1}{M_B} + \frac{1}{M_C}\right) - 2F_{AB,BC}\frac{1}{M_B}\right]\lambda^2 + (F_{AB}F_{BC} - F^2_{AB,BC})\left(\frac{1}{M_AM_B} + \frac{1}{M_AM_C} + \frac{1}{M_BM_C}\right)\lambda = 0 \quad \text{(III-6)}$$

This is a cubic equation, so there are three solutions. One solution is $\lambda = 0$ which is for a translation of the whole molecule in the x direction. This is sometimes referred to as a *nongenuine vibration*. For the other two solutions, this leaves a quadratic equation to solve for λ. If the masses and force constants are known, we can solve for the two values of λ and, thus, the two stretching frequencies ν. However, generally we know the masses and frequencies and desire to solve for the force constants. This requires an *a priori* knowledge of one of the three force constants. Without other information, one would assume that $F_{AB,BC} \approx 0$ since it is second order compared to F_{AB} and F_{BC}.

With a knowledge of the force constants and frequencies, we can go back to Eqs. (III-4) and calculate the relative amplitudes of the various atoms for a given vibrational frequency. Thus, inserting the known values of F_{AB}, F_{BC}, $F_{AB,BC}$, M_A, M_B, M_C, and a given known frequency, λ_k, into the first three equations of (III-4), we have three simultaneous equations with known numerical coefficients

$$\begin{aligned} C_{11}\,\Delta x_A + C_{12}\,\Delta x_B + C_{13}\,\Delta x_C &= 0 \\ C_{21}\,\Delta x_A + C_{22}\,\Delta x_B + C_{23}\,\Delta x_C &= 0 \\ C_{31}\,\Delta x_A + C_{32}\,\Delta x_B + C_{33}\,\Delta x_C &= 0 \end{aligned}$$

We can readily solve these equations for the ratios $\Delta x_B/\Delta x_A$ and $\Delta x_C/\Delta x_A$. These would give the motions of the three atoms for vibration ν_k normalized to $\Delta x_A = 1$. If the AB and BC bonds have appreciably different masses or force constants or both, one value of λ_k will involve primarily stretching of AB and the other primarily stretching of BC. We would then call these the AB and BC stretching frequencies, respectively.

The second three equations of (III-4) can be treated in a similar fashion. Upon expansion of the determinant of the coefficients, we arrive at the secular equation

$$\lambda^3 - F_\alpha\left[\frac{1}{r^2_{AB}M_A} + \frac{1}{r^2_{BC}M_C} + \left(\frac{1}{r_{AB}} + \frac{1}{r_{BC}}\right)^2\frac{1}{M_B}\right]\lambda^2 = 0 \quad \text{(III-7)}$$

This gives us two trivial solutions of $\lambda = 0$ which are for translation of the molecule as a unit in the y direction and rotation about the z axis. The one genuine vibration gives the solution

$$\lambda = F_\alpha\left[\frac{1}{r_{AB}^2 M_A} + \frac{1}{r_{BC}^2 M_C} + \left(\frac{1}{r_{AB}} + \frac{1}{r_{BC}}\right)^2 \frac{1}{M_B}\right] \tag{III-8}$$

An analogous treatment of the Δz_i coordinates [the last three equations of (III-4)] gives an identical solution; thus, we have two bending frequencies which are the same and we say the bending vibration is doubly degenerate.

The same results as obtained in this section can be obtained in a much easier fashion by using internal coordinates and the F and G matrices of Wilson, as described in Sect. 1.6. The purpose of generating the relations through the use of Cartesian coordinates was to give the reader not familiar with vibration theory a feeling for the problem.

Normal Vibrations of Symmetrical Linear *ABA*

The results above can readily be used to derive frequency relations for symmetrical linear ABA, such as CO_2. For the stretching frequencies, one uses Eq. (III-6) and replaces F_{BC} by F_{AB}, and M_C by M_A. After eliminating the trivial solution, $\lambda = 0$, the quadratic equation left is

$$\lambda^2 - \left[F_{AB}\left(\frac{2}{M_A} + \frac{2}{M_B}\right) - F_{AB,AB}\frac{2}{M_B}\right]\lambda + (F_{AB}^2 - F_{AB,AB}^2)\left(\frac{1}{M_A^2} + \frac{2}{M_A M_B}\right) = 0 \tag{III-9}$$

This equation can be rearranged to give

$$\left[\lambda - (F_{AB} + F_{AB,AB})\left(\frac{1}{M_A}\right)\right]\left[\lambda - (F_{AB} - F_{AB,AB})\left(\frac{1}{M_A} + \frac{2}{M_B}\right)\right] = 0$$

Thus we can separate out the two solutions for λ.

$$\lambda_1 = (F_{AB} + F_{AB,AB})\frac{1}{M_A} \tag{III-10}$$

$$\lambda_2 = (F_{AB} - F_{AB,AB})\left(\frac{1}{M_A} + \frac{2}{M_B}\right) \tag{III-11}$$

Substitution of λ_1 in the first three of Eqs. (III-4) yields the solution

$$\Delta x_B = 0, \qquad \Delta x_{A_1} = -\Delta x_{A_2}$$

giving the motion shown below:

$$\leftarrow 0 \text{———} 0 \text{———} 0 \rightarrow$$
$$A_1 \qquad B \qquad A_2$$

This, of course, is known as the symmetric stretching mode.

Substitution of λ_2 gives the solution, $\Delta x_{A_2} = \Delta x_{A_1}$, and $\Delta x_B = -2(M_A/M_B)\Delta x_{A_1}$. The resulting motion is shown below:

$$0\rightarrow \leftarrow 0 \text{———} 0 \rightarrow$$

Since the AB bonds stretch 180° out-of-phase with each other, this is called the antisymmetric stretching vibration.

Substitution of $M_C = M_A$ and $r_{BC} = r_{AB} = r$ in Eq. (III-8) leads to the expression

$$\lambda = \frac{2F_\alpha}{r^2}\left[\frac{1}{M_A} + \frac{2}{M_B}\right] \tag{III-12}$$

which gives the relation between force constant and frequency for the bending vibration of linear symmetrical ABA. Substitution of (III-12) into (III-4) (with $M_C = M_A$ and $r_{AB} = r_{BC} = r$) leads to the amplitudes

$$\Delta y_A = \Delta y_C, \qquad \Delta y_B = \frac{-2M_A}{M_B}(\Delta y_A)$$

describing the motion for the bending vibration, as shown below:

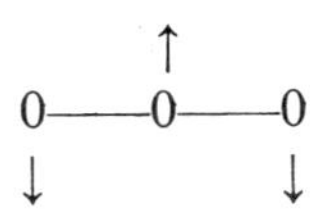

Similar bending in the z direction gives rise to the double degeneracy of this vibrational mode.

Appendix IV

Meaning of Interaction Constants

In Sect. 1.10, the significance of interaction constants was illustrated by introducing an "interaction displacement coordinate," $(l)_k$, by which we mean the displacement of coordinate l which would minimize the energy after unit positive displacement of coordinate k. Another interpretation has sometimes been applied to the interaction constant, F_{kl}; namely, it is said to be the force on coordinate k arising from unit displacement of coordinate l. The purpose of this appendix is to examine such an interpretation. Let us write the potential energy

$$2V = \sum F_i S_i^2 + 2\sum_{i<j} F_{ij} S_i S_j$$

Now consider a unit positive displacement of coordinate k, $S_k = +1$, with all other coordinates remaining unchanged. The potential energy has risen from zero to

$$V = \tfrac{1}{2} F_k$$

The system is no longer at equilibrium. The displacement of coordinate k has disturbed the equilibrium electronic configuration so as to set up forces in the other coordinates. The force on coordinate l, keeping the constraint $S_k = +1$, is now

$$(f_l)_k = -\left(\frac{dV}{dS_l}\right)_{\substack{S_k=+1\\ S_j=0\\ j\neq k}} = -F_{kl}$$

Thus, if F_{kl} is positive, this tells us that a unit positive displacement of k leads to a force tending to decrease the magnitude of coordinate l. This is a useful quantity which gives information about the electronic structure; that is, it would indicate how the electronic forces change for a displacement of a given coordinate. However, there is one disadvantage to the interpretation of this quantity F_{kl}. Its value depends on what internal coordinates are assumed to be important. Consider a bent symmetrical triatomic molecule, XYX. If a General Quadratic Valence Force Field is assumed, the resulting $F_{XY,X'Y}$ is the force on $X'Y$ arising from a unit displacement of XY while the $X'Y$ distance and the angle XYX are held fixed. If a General Quadratic Central Force Field is assumed, the resulting $F_{XY,X'Y}$ is the force on $X'Y$ arising from a unit displacement of XY while the $X'Y$ distance and the XX distance remain fixed. These are different situations and the force constant, $F_{XY,X'Y}$, is different for the two cases. On the other hand, using Eq. (1.10-4) leads to the same "interaction displacement coordinates," $(l)_k$, whether General Quadratic Valence Forces or General Quadratic Central Forces are assumed. Thus the $(l)_k$ are independent of the type of force field assumed as long as it is a General Quadratic Force Field. This means that the $(l)_k$ are more fundamental quantities than the F_{lk}. Furthermore, the quantities $(l)_k$ seem more readily correlated with changes in electronic structure. Thus, if one internal coordinate is constrained to a displaced position, one can expect that this would change the electronic arrangement over the molecule, resulting in new equilibrium values for the other internal coordinates. A knowledge of these changes in equilibrium configuration, specifically the $(l)_k$, will give one an insight into the changes in electronic structure accompanying displacement of one coordinate.

The direct relation of compliance constants to the interaction coordinates, $(l)_k$, as discussed in Section 1.12, shows that compliance constants give a more unique description of interactions than do force constants.

Appendix V

FORCE CONSTANTS OF DIATOMIC MOLECULES

This section lists the harmonic frequencies ω_e, the anharmonicity constants $\omega_e x_e$, and the force constants k_e, for a large number of diatomic molecules. The order of listing poses somewhat of a problem. For quick reference, perhaps alphabetical order is preferable; however, this would disguise some of the periodic trends, so an attempt has been made to list the molecules according to their grouping in the periodic table. Molecules for which the two atoms are in the same group are presented in Table V-1. These are followed in Table V-2 by molecules for which the two atoms are in different groups (except for hydrides). They are grouped according to the more positive atom in the molecule in the order of its group in periodic table. In Table V-3, hydrides are presented, ordered by group in the periodic table for the nonhydrogen atom. Most of the frequencies and anharmonicity constants are taken from *Spectra of Diatomic Molecules* by G. Herzberg (*1*). The original references are given therein and are not repeated here. For the values not taken from Herzberg's book, references are given in these tables.

Those values which bear considerable uncertainty are enclosed in parentheses. For details see Herzberg (*1*). Square brackets around ω_e and k_e indicate the value listed for ω_e is the observed frequency, $\nu(1-0)$, uncorrected for anharmonicity. The isotopic mass numbers are given, where known, as presuperscripts except for atoms having 100% natural abundance of one isotope. The compliance constants in Å $mdyn^{-1}$ are just the reciprocals of the force constants.

REFERENCE

1. G. Herzberg, *Spectra of Diatomic Molecules*, Van Nostrand, Princeton, New Jersey, 1963.

TABLE V-1

HARMONIC FREQUENCIES, ANHARMONICITY CONSTANTS, AND FORCE CONSTANTS FOR SOME DIATOMIC MOLECULES FOR WHICH BOTH ATOMS ARE IN THE SAME GROUP OF THE PERIODIC TABLE

Group	Molecule	ω_e (cm^{-1})	$\omega_e x_e$ (cm^{-1})	k_e (mdyn Å^{-1})
0	He_2^+	[1627.2]		[3.122]
	H_2	4401.22	121.34	5.751[a]
	H_2^+	2297	62	1.566
IA	$^7Li_2$	351.44	2.59	0.255
	Na_2	159.23	0.73	0.172
	NaK	123.29	0.400	0.130
	NaRb	106.64	0.455	0.121
	K_2	92.64	0.354	0.0985
	Rb_2	57.28	0.96	0.082
	RbCs	49.41		0.0748
	Cs_2	41.99	0.08	0.069
IIIA	$^{11}B_2$	1051.3	9.4	3.584
IVA	$^{12}C_2$	1641.35	11.67	9.523
	Si_2	(750)		(4.65)
	Pb_2	256.5	2.96	4.016
VA	$^{14}N_2$	2359.6	14.46	22.966
	$^{14}N_2^+$	2207.19	16.14	20.095
	$^{14}N^{31}P$	1337.24	6.983	10.160
	P_2	780.43	2.80	5.557
	$^{14}N^{75}As$	1068.0	5.36	7.928
	^{14}NSb	942.0	5.6	6.566
	As_2	429.44	1.12	4.069
	As_2^+	(314.8)	(1.25)	(2.187)
	Sb_2	269.85	0.59	2.612
	SbBi	220.0	0.50	2.194
	Bi_2	172.71	0.32	1.836
VIA	$^{16}O_2$	1580.36	12.07	11.768
	$^{16}O_2^+$	1876.4	16.53	16.589
	$^{16}O^{32}S$	1123.73	6.116	7.932
	$^{32}S_2$	725.68	2.85	4.960
	^{16}OSe	907.1	4.61	6.448
	$^{80}Se_2$	391.77	1.06	3.613
	^{16}OTe	796.0	3.50	5.305
	Te_2	251	0.55	2.37
VIIA	F_2	[892.1]		[4.454]
	^{35}ClF	793.2	9.9	4.562
	^{79}BrF	671	3	4.071
	$^{35}Cl_2$	564.9	4.0	3.287
	$^{35}Cl_2^+$	(645.3)	(2.9)	(4.289)
	BrCl	442.5	1.5	2.832[b]
	$^{79}Br_2$	325.37	1.098	2.461[c]
	$^{81}Br_2$	321.29	1.064	2.460[c]
	$^{79}Br^{81}Br$	323.2	1.07	2.459
	$I^{35}Cl$	384.18	1.465	2.383
	$I^{79}Br$	268.4	0.78	2.064
	I_2	214.57	0.61	1.721
IIB	Hg_2	(36)		(0.077)

[a] U. Fink, T. A. Wiggins, and D. H. Rank, *J. Mol. Spectry.*, **18**, 384 (1965).
[b] W. V. F. Brooks and B. L. Crawford, Jr., *J. Chem. Phys.*, **23**, 363 (1955).
[c] J. A. Horsley and R. F. Barrow, *Trans. Faraday Soc.*, **63**, 32 (1967).

TABLE V-2

HARMONIC FREQUENCIES, ANHARMONICITY CONSTANTS, AND FORCE CONSTANTS FOR SOME DIATOMIC MOLECULES WITH THE TWO ATOMS IN DIFFERENT GROUPS OF THE PERIODIC TABLE (HYDRIDES NOT INCLUDED)

Group[a]	Molecule	ω_e (cm^{-1})	$\omega_e x_e$ (cm^{-1})	k_e (mdyn $Å^{-1}$)
IA	${}^{7}LiF$	906.2	7.90	2.479[b]
	${}^{7}LiCl$	641	4.2	1.417[b]
	${}^{7}LiBr$	563	3.53	1.204[b]
	${}^{7}LiI$	498	3.39	0.971[b]
	NaF	536.1	3.83	1.761[c]
	NaCl	366	2.05	1.100[d]
	NaBr	302	1.50	0.959[d]
	NaI	258	1.08	0.763[d]
	KF	426	2.4	1.37[e]
	KCl	281	1.36	0.864[d]
	KBr	213	0.80	0.702[d]
	RbF	376	1.9	1.30[e]
	RbCl	228	0.92	0.767[d]
	CsF	353	1.7	1.22[e]
	CsCl	209	0.75	0.719[d]
IIA	$Be^{16}O$	1478.32	11.83	7.512
	BeF	1265.6	9.12	5.768
	$Be^{35}Cl$	846.58	5.11	3.026
	${}^{24}Mg^{16}O$	785.1	5.18	3.485
	MgS	525.2	2.93	2.247
	${}^{24}MgF$	717.6	3.84	3.216
	${}^{24}Mg^{35}Cl$	465.4	2.05	1.815
	${}^{24}Mg^{79}Br$	373.8	1.34	1.514
	${}^{40}Ca^{16}O$	650	6.6	2.843
	${}^{40}CaF$	587.1	2.74	2.615
	$Ca^{35}Cl$	369.8	1.31	1.505
	$Ca^{79}Br$	285.3	0.86	1.275
	CaI	242.0	0.64	1.051
	$Sr^{16}O$	653.5	4.0	3.402
	SrF	500.1	2.21	2.301
	$Sr^{35}Cl$	302.3	0.95	1.346
	$Sr^{79}Br$	216.5	0.51	1.147
	SrI	173.9	0.42	0.924
	$Ba^{16}O$	669.8	2.05	3.787
	BaF	468.9	1.79	2.162
	${}^{138}Ba^{35}Cl$	279.3	0.89	1.282
	$Ba^{79}Br$	193.8	0.42	1.109
IIIA	${}^{11}B^{14}N$	1514.6	12.3	8.330
	${}^{11}B^{16}O$	1885.44	11.769	13.657
	${}^{11}BF$	1399.8	11.3	8.047
	${}^{11}B^{35}Cl$	839.12	5.11	3.473
	${}^{11}B^{79}Br$	684.31	3.52	2.665
	$Al^{16}O$	978.2	7.12	5.661
	AlF	814.5	8.1	4.357
	$Al^{35}Cl$	481.30	1.95	2.079
	$Al^{79}Br$	378.0	1.28	1.693
	AlI	316.1	1.0	1.310
	$Ga^{16}O$	767.69	6.34	4.517
	$Ga^{35}Cl$	365.0	1.1	1.821
	$Ga^{81}Br$	263.0	0.81	1.517
	GaI	216.4	0.5	1.232
	$In^{16}O$	703.09	3.17	4.088
	${}^{115}In^{35}Cl$	317.4	1.01	1.591
	${}^{115}In^{81}Br$	221.0	0.65	1.366
	${}^{115}InI$	177.1	0.4	1.114
	TlF	475.00	1.89	2.311
	$Tl^{35}Cl$	287.47	1.24	1.454
	$Tl^{81}Br$	192.1	0.39	1.260
	TlI	(150)		(1.038)
IVA	${}^{12}C^{14}N$	2068.71	13.14	16.293
	${}^{12}C^{14}N^{-}$	2106 ±2	13 ±2	16.89 ±0.04[f]
	${}^{12}CP$	1239.67	6.86	7.831
	${}^{12}C^{16}O$	2169.82	13.29	19.017[g]
	${}^{12}C^{16}O^{+}$	2214.24	15.16	19.803
	${}^{12}C^{32}S$	1285.1	6.5	8.489
	${}^{12}CSe$	1036.0	4.8	6.587
	${}^{12}C^{35}Cl$	846	1.0	3.767
	${}^{28}Si^{14}N$	1151.68	6.560	7.293

[a] This is the group of the periodic table in which the first atom (more positive atom) lies.

[b] W. Klemperer, W. G. Norris, A. Büchler, and A. G. Emslie, *J. Chem. Phys.*, **33,** 1534 (1960).

[c] S. E. Veazey and W. Gordy, *Phys. Rev.*, **138,** A 1303 (1965).

[d] S. A. Rice and W. Klemperer, *J. Chem. Phys.*, **27,** 573 (1957).

[e] V. I. Baikov and K. P. Vasilevskii, *Opt. Spectry.*, **22,** 364 (1967).

[f] In aqueous solution, $\nu_0 = 2080\ cm^{-1}$. [R. A. Penneman and L. H. Jones, *J. Chem. Phys.*, **24,** 293 (1956).]

[g] D. H. Rank, A. G. St. Pierre, and T. A. Wiggins, *J. Mol. Spectry.*, **18,** 418 (1965).

TABLE V-2 (contd.)

Group	Molecule	ω_e (cm^{-1})	$\omega_e x_e$ (cm^{-1})	k_e (mdyn Å^{-1})
	^{28}Si^{16}O	1242.03	6.047	9.249
	^{28}Si^{32}S	749.5	2.56	4.938
	^{28}SiSe	580.0	1.78	4.094
	^{28}SiTe	481.2	1.30	3.130
	^{74}Ge^{16}O	985.7	4.30	7.527
	^{74}Ge^{32}S	575.8	1.80	4.360
	^{74}Ge^{80}Se	406.8	1.2	3.744
	^{74}Ge^{130}Te	323.4	1.0	2.903
	GeF	665.2	2.79	3.926
	^{74}Ge^{35}Cl	407.6	1.36	2.324
	GeBr	296.6	0.9	1.972
	Sn^{16}O	822.4	3.73	5.617
IVA	SnS	487.68	1.34	3.537
contd.	SnSe	331.2	0.736	3.064
	SnTe	259.5	0.50	2.440
	SnF	582.9	2.69	3.278
	Sn^{35}Cl	352.5	1.06	1.977
	SnBr	247.7	0.62	1.726
	Pb^{16}O	721.8	3.70	4.558
	^{208}Pb^{32}S	428.14	1.201	2.993
	PbSe	277.6	0.51	2.596
	PbTe	211.8	0.12	2.087
	PbF	507.2	2.30	2.638
	Pb^{35}Cl	303.8	0.88	1.627
	Pb^{79}Br	207.5	0.50	1.450
	PbI	160.5	0.25	1.194
	^{14}N^{16}O	1903.86	13.97	15.944[h]
	^{15}N^{16}O	1869.8	13.47	15.944[h,i]
	^{14}N^{32}S	1220.0	7.75	8.539
	^{14}NBr	693	5.0	3.371
	P^{16}O	1230.64	6.52	9.412
	As^{16}O	967.4	5.3	7.267
VA	Sb^{16}O	817.2	5.30	5.562
	SbF	614.2	2.77	3.653
	Sb^{35}Cl	369.0	0.92	2.179
	Bi^{16}O	702.1	5.20	4.315
	BiF	510.7	2.05	2.676
	Bi^{35}Cl	308.0	0.96	1.674
	Bi^{79}Br	209.34	0.468	1.479
VA	BiI	163.9	0.31	1.250
	^{16}OCl	(780)		(3.951)
VIA	^{16}OBr	713	7	3.99
	^{16}OI	(687)	(5)	(3.95)
	Cu^{16}O	628	3	2.969
	^{63}CuF	622.7	3.95	3.333
	^{63}Cu^{35}Cl	416.9	1.57	2.281
	^{63}Cu^{79}Br	314.10	0.865	2.035
	^{63}CuI	264.8	0.71	1.738
IB	Ag^{16}O	493.2	4.10	1.996
	^{107}Ag^{35}Cl	343.6	1.163	1.833
	^{109}Ag^{81}Br	247.72	0.680	1.678
	^{107}AgI	206.18	0.433	1.453
	Au^{35}Cl	382.8	1.30	2.564
	ZnF	(630)	(3.5)	(3.442)
	Zn^{35}Cl	390.5	1.55	2.047
	ZnBr	(220)		(1.026)
	^{64}ZnI	223.4	0.75	1.250
	CdF	(535)		[2.74]
	Cd^{35}Cl	330.5	1.2	1.716
IIB	CdBr	230.0	0.50	1.456
	CdI	178.5	0.625	1.119
	HgTl	26.9	0.69	0.043
	HgF	490.8	4.05	2.463
	Hg^{35}Cl	292.61	1.603	1.502
	^{202}Hg^{81}Br	186.25	0.975	1.181
	HgI	125.6	1.09	0.722
	^{45}Sc^{16}O	971.55	3.95	6.561
	^{89}Y^{16}O	852.5	2.45	5.804
IIIB	^{139}La^{16}O	811.6	2.23	6.116
and	Ce^{16}O	865.0	2.99	6.328
Lanth-	^{141}Pr^{16}O	818.9	1.20	5.675
anides	Gd^{16}O	841.0	3.70	6.049
	Lu^{16}O	841.66	4.07	6.116
	YbCl	293.61	1.23	1.495
IVB	^{48}Ti^{16}O	1008.4	4.61	7.186

[h] For NO the reported ω_e is the average for the doublet ground state.

[i] For ^{15}N^{16}O, the ω_e and $\omega_e x_e$ are from infrared studies of W. H. Fletcher and G. M. Begun, *J. Chem. Phys.*, **27**, 579 (1957).

TABLE V-2 (contd.)

Group	Molecule	ω_e (cm^{-1})	$\omega_e x_e$ (cm^{-1})	k_e (mdyn Å^{-1})	Group	Molecule	ω_e (cm^{-1})	$\omega_e x_e$ (cm^{-1})	k_e (mdyn Å^{-1})
IVB contd.	$^{48}Ti^{35}Cl$	456.4	6.3	2.482	VIIB contd.	^{55}MnF	618.8	3.01	3.185
	$^{90}Zr^{16}O$	936.6	3.45	7.018		$^{55}Mn^{35}Cl$	384.9	1.4	1.865
						$^{55}MnBr$	289.7	0.9	1.610
VB	$V^{16}O$	1012.7	4.9	7.355					
					VIII	$Fe^{16}O$	880	5	5.673
VIB	$Cr^{16}O$	898.8	6.5	5.822		$Fe^{35}Cl$	406.6	1.2	2.095
						CoCl	421.2	0.74	2.314
VIIB	$^{55}Mn^{16}O$	840.7	4.89	5.159		NiCl	419.2	1.04	2.288

TABLE V-3

HARMONIC FREQUENCIES, ANHARMONICITY CONSTANTS, AND FORCE CONSTANTS FOR SOME DIATOMIC HYDRIDES

Group	Molecule	ω_e (cm^{-1})	$\omega_e x_e$ (cm^{-1})	k_e (mdyn $Å^{-1}$)
IA	^{7}LiH	1405.65	23.2	1.026
	^{7}LiD	1055.12	13.23	1.026
	^{23}NaH	1172.2	19.72	0.782
	^{39}KH	985.0	14.65	0.562
	^{39}KD	705.95	7.6	0.562
	RbH	936.77	14.15	0.515
	CsH	890.7	12.6	0.468
IIA	BeH	2058.6	35.5	2.263
	BeH^+	2221.7	39.79	2.635
	BeD^+	1647.64	21.85	2.633
	MgH	1495.7	31.5	1.274
	MgD	1077.76	16.09	1.272
	MgH^+	1695.3	30.2	1.637
	MgD^+	1226.6	16.30	1.647
	^{40}CaH	1299	19.5	0.977
	SrH	1206.2	17.0	0.854
	BaH	1172	16	0.810
IIIA	^{11}BH	(2366)	(49)	(3.045)
	^{11}BD	(1780)	(27)	(3.178)
	$^{11}BH^+$	(2435)		(3.225)
	AlH	1682.57	29.15	1.621
	AlD	1211.95	15.14	1.622
	AlH^+	(1610)		1.484
	TlH	1390.7	22.7	1.143
IVA	^{12}CH	2861.6	64.3	4.485
	^{12}CD	2101.0	34.7	4.485
	$^{12}CH^+$	[2739.54]		[4.110]
	^{28}SiH	2042.47	35.67	2.377[a]
	SnH	(1580)		(1.470)
	PbH	1564.1	29.75	1.445
VA	^{14}NH	(3300)		(6.032)
	PH	(2380)		(3.258)
	BiH	1698.9	31.6	1.705
	BiD	1205.5	16.1	1.708
VIA	^{16}OH	3735.21	82.81	7.793
	^{16}OD	2720.9	44.2	7.802
	$^{16}OH^+$	[2955]		[4.877]
VIIA	HF	4138.73	90.05	9.658[b]
	DF	2998.19	45.76	9.644[c]
	TF	2508.5	32.5	9.649[d]
	H^{35}Cl	2990.97	52.85	5.163[e]
	D^{35}Cl	2145.16	27.18	5.163[e]
	$H^{35}Cl^+$	2675.4	53.5	4.131
	HBr	2649.67	45.21	4.117
	DBr	1885.04	22.72	4.113[f]
	HI	2309.53	39.73	3.142
	DI	1640.14	20.16	3.142[g]
IB	^{63}CuH	1940.4	37.0	2.201
	^{63}CuD	1384.38	19.14	2.204
	$^{63}CuH^+$	(1874)		(2.053)
	AgH	1760.0	34.05	1.823
	AuH	2305.01	43.12	3.139
	AuD	1634.98	21.66	3.140
IIB	ZnH	1607.6	55.14	1.511
	ZnH^+	1916	39	2.147
	ZnD^+	1364.8	19.8	2.144
	CdH	1430.7	46.3	1.205
	CdH^+	1775.4	37.3	1.855
	CdD^+	1262.5	19.01	1.858
	HgH	1387.09	83.01	1.137
	HgD	995.15	49.93	1.163?
	HgH^+	2033.87	46.16	2.444
	HgD^+	1442.15	23.24	2.443
VIIB	^{55}MnH	[1490.58]		[1.296]
VIII	CoH	(1890)		(2.086)
	NiH	[1926.6]		[2.167]

[a] A. E. Douglas, *Can. J. Phys.*, **35,** 71 (1957).

[b] D. E. Mann, B. A. Thrush, D. R. Lide, Jr., J. J. Ball, and N. Acquista, *J. Chem. Phys.*, **34,** 420 (1961).

[c] R. N. Spanbauer, K. N. Rao and L. H. Jones, *J. Mol. Spectry.*, **16,** 100 (1965).

[d] L. H. Jones and M. Goldblatt, *J. Mol. Spectry.*, **1,** 43 (1957).

[e] D. H. Rank, D. P. Eastman, B. S. Rao, and T. A. Wiggins, *J. Opt. Soc. Am.*, **52,** 1 (1962).

[f] F. L. Keller and A. H. Nielsen, *J. Chem. Phys.*, **22,** 294 (1954).

[g] L. H. Jones, *J. Mol. Spectry.*, **1,** 179 (1957).

Appendix VI

Normal Coordinate Treatment and Meaning of Force Constants for Planar $MX_3(D_{3h}$ Symmetry)

The internal displacement coordinates used here for MX_3 are defined as r_i (the change in bond length MX_i), α_{ij} (the change in angle X_iMX_j), and π (the change in angle bond MX_1 makes with the plane of X_2MX_3). The potential energy expression for a general quadratic valence force field is then

$$\begin{aligned}2V = {} & f_\alpha(\alpha_{12} + \alpha_{13} + \alpha_{23}) + F_r(r_1^2 + r_2^2 + r_3^2) \\ & + 2F'_{rr}(r_1r_2 + r_1r_3 + r_2r_3) + F_\alpha(\alpha_{12}^2 + \alpha_{13}^2 + \alpha_{23}^2) \\ & + 2F'_{\alpha\alpha}(\alpha_{12}\alpha_{13} + \alpha_{12}\alpha_{23} + \alpha_{13}\alpha_{23}) \\ & + 2F'_{r\alpha}[r_1(\alpha_{12} + \alpha_{13}) + r_2(\alpha_{12} + \alpha_{23}) + r_3(\alpha_{13} + \alpha_{23})] \\ & + 2F''_{r\alpha}[r_1\alpha_{23} + r_2\alpha_{13} + r_3\alpha_{12}] + F_\pi\pi^2 \qquad \text{(VI-1)}\end{aligned}$$

As discussed in Sect. 1.10, the first-order term must be included because the α_{ij} are not independent. The redundancy condition to the second order as developed by the method of Cihla and Pliva (*1*) is given by

$$\begin{aligned}\alpha_{12} + \alpha_{13} + \alpha_{23} + \frac{1}{2\sqrt{3}}[\alpha_{12}^2 + \alpha_{13}^2 + \alpha_{23}^2 + 2\pi^2 \\ + 2\alpha_{12}\alpha_{13} + 2\alpha_{12}\alpha_{23} + 2\alpha_{13}\alpha_{23}] = 0 \qquad \text{(VI-2)}\end{aligned}$$

By substitution of (VI-2) in (VI-1), the first order term can be eliminated, whereby the F_α, $F'_{\alpha\alpha}$, and F_π coefficients become $\bar{F}_\alpha = F_\alpha - (1/2\sqrt{3})f_\alpha$, $\bar{F}'_{\alpha\alpha} = F'_{\alpha\alpha} - (1/2\sqrt{3})f_\alpha$, and $\bar{F}_\pi = F_\pi - (1/\sqrt{3})f_\alpha$.

Considering the first-order redundancy, it is seen that in Eq. (VI-1) the coordinates following $F'_{r\alpha}$ are just the negative of those following $F''_{r\alpha}$. Thus, we can simplify Eq. (VI-1) somewhat (letting $H_{r\alpha} = F''_{r\alpha} - F'_{r\alpha}$) to

$$\begin{aligned}2V = {} & F_r(r_1^2 + r_2^2 + r_3^2) + 2F'_{rr}(r_1r_2 + r_1r_3 + r_2r_3) \\ & + \bar{F}_\alpha(\alpha_{12}^2 + \alpha_{13}^2 + \alpha_{23}^2) \\ & + 2\bar{F}'_{\alpha\alpha}(\alpha_{12}\alpha_{13} + \alpha_{12}\alpha_{23} + \alpha_{13}\alpha_{23}) \\ & + 2H_{r\alpha}[r_1\alpha_{23} + r_2\alpha_{13} + r_3\alpha_{12}] + \bar{F}_\pi\pi^2 \qquad \text{(VI-3)}\end{aligned}$$

Furthermore, we can eliminate the redundancy in the $\alpha\alpha$ terms by letting $\alpha_{23} = -\alpha_{12} - \alpha_{13}$ and we arrive at

$$\begin{aligned}2V &= F_r(r_1^2 + r_2^2 + r_3^2) + 2F'_{rr}(r_1r_2 + r_1r_3 + r_2r_3) \\ &\quad + 2(F_\alpha - F'_{\alpha\alpha})(\alpha_{12}^2 + \alpha_{13}^2 + \alpha_{12}\alpha_{13}) \\ &\quad + 2H_{r\alpha}[r_1\alpha_{23} + r_2\alpha_{13} + r_3\alpha_{12}] + \bar{F}_\pi\pi^2 \end{aligned} \tag{VI-4}$$

By the methods outlined in Sect. 1.10, we can now investigate the physical meaning of the various force constants. The results are

$F_r = (\partial^2 V/\partial r^2)_0 =$ the force resisting distortion of an MX bond, or the bond strength at the equilibrium electronic configuration

$F_\pi =$ the force resisting bending of one of the bonds out of the plane

$$F_{rr} = \frac{-F_r(r_2)_{r_1} - H_{r\alpha}(\alpha_{12})_{r_1}}{1 + (r_2)_{r_1}} \tag{VI-5}$$

$$H_{r\alpha} = \frac{3(F_\alpha - F_{\alpha\alpha})(\alpha_{12})_{r_1}}{1 - (r_2)_{r_1}} = 2(F_r - F'_{rr})(r_1)_{\alpha_{12}} = -(F_r - F'_{rr})(r_3)_{\alpha_{12}} \tag{VI-6}$$

$(r_2)_{r_1} =$ displacement of r_2 which minimizes energy after a small unit positive displacement of r_1

$(\alpha_{12})_{r_1} =$ displacement of α_{12} which minimizes energy after a small unit positive displacement of r_1

$(r_1)_{\alpha_{12}} =$ displacement of r_1 which minimizes energy after a small unit positive displacement of α_{12}

Perhaps the best physical meaning of the bending force constant is obtained from the two expressions

$$F_\alpha - F'_{\alpha\alpha} = \frac{2}{3}\left[\frac{\partial^2 V}{\partial \alpha_{12}^2}\right]_{r_i=0,\alpha_{13}=\alpha_{23}} \tag{VI-7}$$

$$F_\alpha - F'_{\alpha\alpha} = \frac{1}{2}\left[\frac{\partial^2 V}{\partial \alpha_{12}^2}\right]_{r_i=0,\alpha_{13}=-\alpha_{12}} \tag{VI-8}$$

Thus, $F_\alpha - F'_{\alpha\alpha}$ is equal to (2/3) the force constant resisting distortion of type A, or $\frac{1}{2}$ the force constant resisting distortion of type B, in Fig. VI-1. A similar treatment for illustration of square planar MX_4 has been given by Jones (*2*).

In most of the normal coordinate treatments of MX_3, the G matrices used are incorrect; for example, $G_{33}^{E'}$ and $G_{66}^{A_2''}$ (which should be called

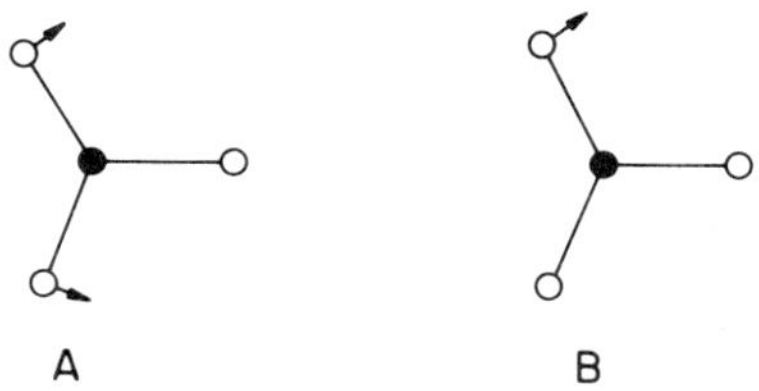

Fig. VI-1 Force constant for distortion $A = 3(F_\alpha - F'_{\alpha\alpha})$; force constant for distortion $B = 2(F_\alpha - F'_{\alpha\alpha})$.

$G_{44}^{E'}$ and $G_{22}^{A_2''}$) of (*1*), Chapter 4, are incorrect. Konaka et al. [in (*6*), Chapter 4] have given the correct expression for $G_{22}^{A_2''}$. They are best obtained by the method of Wilson et al. (*3*). The correct relations among frequencies and force constants are given below, for planar MX_3 (of D_{3h} symmetry).

$$A_1' \qquad \lambda_1 = \mu_X F_{11}$$

$$A_2'' \qquad \lambda_2 = (3\mu_X + 9\mu_M)F_{22}$$

$$E' \qquad \lambda_3 + \lambda_4 = (\tfrac{3}{2}\mu_M + \mu_X)F_{33} + (3\mu_X + \tfrac{9}{2}\mu_M)\rho_0^2 F_{44} + 3\sqrt{3}\,\mu_M \rho_0 F_{34}$$

$$\lambda_3\lambda_4 = (3\mu_X^2 + 9\mu_X\mu_M)\rho_0^2(F_{33}F_{44} - F_{34}^2)$$

$$\lambda_i = 4\pi^2\nu_i^2 c^2$$

where ν_i is frequency in cm^{-1}, c is velocity of light in cm sec^{-1},

$$\mu_j = \frac{1}{M_j}\ \text{gms}^{-1}, \qquad \rho_0 = \frac{1}{r_{MX}}\ \text{Å}^{-1}$$

with r_{MX} the equilibrium MX distance.

The force constants are symmetry force constants defined as:

$$F_{11} = F_r + 2F_{rr} \qquad \text{mdyn Å}^{-1}$$

$$F_{22} = \bar{F}_\pi \qquad \text{mdyn Å rad}^{-2}$$

$$F_{33} = F_r - F_{rr} \qquad \text{mdyn Å}^{-1}$$

$$F_{44} = F_\alpha - F'_{\alpha\alpha} \qquad \text{mdyn Å rad}^{-2}$$

$$F_{34} = F''_{r\alpha} - F'_{r\alpha} \qquad \text{mdyn rad}^{-1} = H_{r\alpha}$$

REFERENCES

1. Z. Cihla and J. Pliva, *Collection Czech. Chem. Commun.*, **26**, 1903 (1961).
2. L. H. Jones, *Coordination Chem. Rev.*, **1**, 351 (1966).
3. E. B. Wilson, J. C. Decius, and P. C. Cross, *Molecular Vibrations*, McGraw-Hill, New York, 1955, pp. 58–59.

Appendix VII

Urey–Bradley Treatment of Planar $XY_3(D_{3h})$

The Urey–Bradley potential function for planar XY_3 of D_{3h} symmetry is given by

$$
\begin{aligned}
V = {} & f_r \sum_i \Delta r_i + \tfrac{1}{2} F_r \sum_i (\Delta r_i)^2 \\
& + f_\alpha \sum_{i<j} \Delta\alpha_{ij} + \tfrac{1}{2} F_\alpha \sum_{i<j} (\Delta\alpha_{ij})^2 \\
& + f_q \sum_{i<j} \Delta q_{ij} + \tfrac{1}{2} F_q \sum_{i<j} (\Delta q_{ij})^2 \\
& + \tfrac{1}{2} F_\pi \Delta\pi^2
\end{aligned}
\qquad \text{(VII-1)}
$$

with

Δr_i	change in distance of XY_i
$\Delta\alpha_{ij}$	change in angle Y_iXY_j
Δq_{ij}	change in distance Y_iY_j
$\Delta\pi$	change in angle XY_1 makes with plane of Y_2XY_3

The first-order terms are necessary because of redundancies among the various α_{ij}, and among the r_i, the q_i, and the α_{ij}. By eliminating these redundancies, the number of force constants is reduced, as follows. According to the law of cosines,

$$q_{ij}^2 = r_i^2 + r_j^2 - 2r_ir_j \cos \alpha_{ij}$$

By substituting for instantaneous values of the coordinates

$$
\begin{aligned}
q_{ij} &= q_0 + \Delta q_{ij} \\
r_i &= r_0 + \Delta r_i \\
\alpha_{ij} &= \alpha_0 + \Delta\alpha_{ij}
\end{aligned}
$$

an expression for Δq_{ij} is obtained in terms of the Δr_i and $\Delta\alpha_{ij}$. Shimanouchi (*1*) gives the general expression for Δq_{ij} to the second order.

For the case of XY_3D_{3h}, this reduces to

$$\Delta q_{ij} = \frac{\sqrt{3}}{2}(\Delta r_i + \Delta r_j) + \frac{1}{2} r_0 \,\Delta\alpha_{ij} + \frac{1}{8\sqrt{3}\, r_0}(\Delta r_i^2 + \Delta r_j^2)$$
$$- \frac{3}{8} r_0^2(\Delta\alpha_{ij})^2 + \frac{\sqrt{3}}{4} r_0 \,\Delta\alpha_{ij}(\Delta r_i + \Delta r_j) - \frac{1}{4}\Delta r_i \,\Delta r_j$$

If we use this expression to eliminate Δq_{ij} from (VII-1), the potential energy is expressed as

$$V = (f_r + \sqrt{3} f_q) \sum_i \Delta r_i + \left(\frac{1}{2} F_r + \frac{f_q}{4\sqrt{3}\, r_0} + \frac{3}{4} F_q\right) \sum_i (\Delta r_i)^2$$
$$+ \left[f_\alpha + \frac{1}{2} f_q r_0\right] \sum_{i<j} \Delta\alpha_{ij} + \left[\frac{1}{2} F_\alpha - \frac{\sqrt{3}}{8} f_q r_0 + \frac{1}{8} F_q r_0^2\right] \sum_{i<j} (\Delta\alpha_{ij})^2$$
$$+ \left[\frac{1}{4} f_q + \frac{\sqrt{3}}{4} F_q r_0\right] \sum_{i<j} (\Delta\alpha_{ij})(\Delta r_i + \Delta r_j)$$
$$+ \left[\frac{3}{4} F_q - \frac{f_q}{4\sqrt{3}\, r_0}\right] \sum_{i<j} r_i r_j + \frac{1}{2} F_\pi (\Delta\pi)^2 \qquad \text{(VII-2)}$$

The Δr_i are now independent so $(dV/d\Delta r)_0 = 0$ and $f_r = -\sqrt{3} f_q$. The redundancy condition to the second order is given for $\Delta\alpha_{ij}$ in Appendix VI. By substitution in (VII-2), we can eliminate the first-order term in $\Delta\alpha_{ij}$. The resulting potential energy expression is

$$V = \left(\frac{1}{2} F_r + \frac{f_q}{4\sqrt{3}\, r_0} + \frac{3}{4} F_q\right) \sum_i (\Delta r_i)^2$$
$$+ \left[\frac{1}{2} F_\alpha - \rho - \frac{\sqrt{3}}{8} f_q r_0 + \frac{1}{8} F_q r_0^2\right] \sum_{i<j} (\Delta\alpha_{ij})^2$$
$$- 2\rho \sum_{i\neq j\neq k} \Delta\alpha_{ij}\, \Delta\alpha_{ik}$$
$$+ \left[\frac{1}{4} f_q + \frac{\sqrt{3}}{4} F_q r_0\right] \sum_{i<j} (\Delta\alpha_{ij})(\Delta r_i + \Delta r_j)$$
$$+ \left(\frac{3}{4} F_q - \frac{f_q}{4\sqrt{3}\, r_0}\right) \sum_{i<j} \Delta r_i \,\Delta r_j + \frac{1}{2} \bar{F}_\pi (\Delta\pi)^2 \qquad \text{(VII-3)}$$

$$\rho = f_\alpha + \tfrac{1}{2} f_q r_0 \qquad \text{and} \qquad \bar{F}_\pi = F_\pi - \rho$$

This expression can now be compared directly with Eq. (VI-3) to relate the Urey–Bradley force constants to the general quadratic valence force

TABLE VII-1

SYMMETRY FORCE CONSTANTS OF XY_3 OF D_{3h} SYMMETRY

		GQVFF	UBFF[a]
F_{11}	(A_1')	$F_r + 2F_{rr}$	$F_r + 3F_q$
F_{22}	(A_2'')	$\bar{F}_\pi$	$\bar{F}_\tau$
F_{33}	(E')	$F_r - F_{rr}$	$F_r + \frac{\sqrt{3}}{4}\frac{f_q}{r_0} + \frac{3}{4}F_q$
F_{34}	(E')	$F''_{r\alpha} - F'_{r\alpha}$	$-\left(\frac{1}{4}f_q + \frac{\sqrt{3}}{4}r_0F_q\right)$
F_{44}	(E')	$F_\alpha - F'_{\alpha\alpha}$	$F_\alpha - \frac{\sqrt{3}}{4}r_0f_q + \frac{1}{4}F_qr_0^2$

[a] These expressions are equivalent to those of Meisingseth (*2*) with different nomenclature.

constants. This is done in Table VII-1. Considering $F'_{r\alpha} - F''_{r\alpha}$ and $F_\alpha - F'_{\alpha\alpha}$ as two force constants, there are five force constants to determine for the GQVFF. There are also five to determine for the UBFF. However, it is generally accepted (*1*, *2*) that $(f_q/q_0) \approx -0.1F_q$ and with this approximation one can solve for the four UBFF constants (F_r, F_α, F_q, and F_π) from the four observed frequencies.

REFERENCES

1. T. Shimanouchi, *J. Chem. Phys.*, **17**, 245 (1949).
2. E. Meisingseth, *Acta Chem. Scand.*, **16**, 1601 (1962).

Appendix VIII

NORMAL COORDINATE TREATMENT OF MX_6 OF O_h SYMMETRY

The internal displacement coordinates are defined by Fig. VIII-1. The potential energy expression is rather cumbersome because of the several different types of bend–bend interactions. It is given below in an abbreviated form;

$$\begin{aligned} 2V = {} & f_\alpha \sum_{ij} \alpha_{ij} + F_r \sum r_i^2 + 2F''_{rr}(r_1 r_2 + \cdots) \\ & + 2F'_{rr}(r_1 r_6 + \cdots) + F_\alpha \sum \alpha_{ij}^2 \\ & + 2F'_{\alpha\alpha}(\alpha_{12}\alpha_{13} + \cdots) + 2F''_{\alpha\alpha}(\alpha_{12}\alpha_{46} + \cdots) \\ & + 2F'''_{\alpha\alpha}(\alpha_{12}\alpha_{36} + \cdots) + 2F''''_{\alpha\alpha}(\alpha_{12}\alpha_{14} + \cdots) \\ & + 2F'_{r\alpha}(r_1\alpha_{12} + \cdots) + 2F''_{r\alpha}(r_1\alpha_{26} + \cdots) \end{aligned} \qquad \text{(VIII-1)}$$

with r_1 the change in bond length MX and α_{ij} the change in angle X_iMX_j.

In this form we can see what interactions the various interaction constants refer to. The first-order term, $f\Sigma\alpha_{ij}$, must be included because the α_{ij} are not independent (see Sect. 1.10). The redundancy condition to the second order is

$$\overset{\text{12 terms}}{\sum_{i,j=1} \alpha_{ij}} = - \overset{\text{24 terms}}{\sum_{i,j,k} \alpha_{ik}\alpha_{kj}} \qquad \text{(VIII-2)}$$

with i and j not opposite; i and k not opposite; j and k not opposite; $i < j$; $i, j \neq k$.

The first-order term in (VIII-1) is eliminated by substituting the expression (VIII-2). The only effect is to change the coefficient of the terms such as $\alpha_{12}\alpha_{13}$ from $F'_{\alpha\alpha}$ to $\overline{F'_{\alpha\alpha}} = F'_{\alpha\alpha} - \frac{1}{2}f_\alpha$.

By methods similar to those in Sect. 1.10 for tetrahedral XY_4, we can investigate the physical meaning of the interaction constants defined in

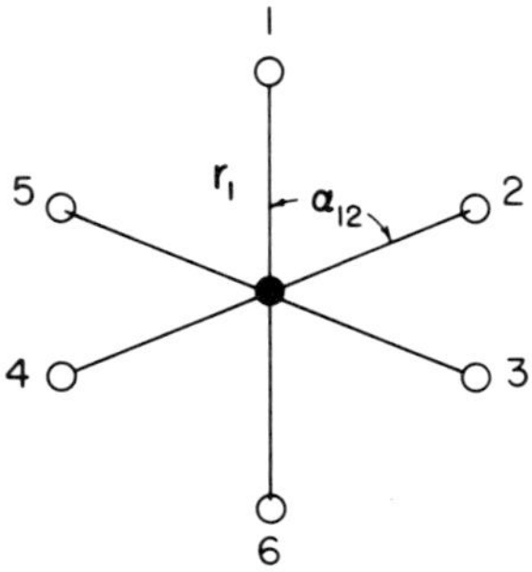

Fig. VIII-1 Internal coordinate description of octahedral XY_6.

(VIII-1). The results are

$$F''_{rr} = \frac{-(F_r + F'_{rr})(r_2)_{r_1}}{1 + 2(r_2)_{r_1} + (r_6)_{r_1}} \tag{VIII-3}$$

$$F'_{rr} = -F_r(r_6)_{r_1} - 4F''_{rr}(r_2)_{r_1} + 4H'_{r\alpha}(\alpha_{12})_{r_1} \tag{VIII-4}$$

$$H'_{r\alpha} = F'_{r\alpha} - F''_{r\alpha} = \frac{-2[(F_\alpha - F''_{\alpha\alpha}) + 2(F'_{\alpha\alpha} - F'''_{\alpha\alpha})](\alpha_{12})_{r_1}}{1 - (r_6)_{r_1}} \tag{VIII-5}$$

$$H'_{r\alpha} = \frac{-(F_r - F'_{rr})(r_1)_{\alpha_{12}}}{1 + (\alpha_{14})_{\alpha_{12}} + 2(\alpha_{13})_{\alpha_{12}}} \tag{VIII-6}$$

$$\overline{F'_{\alpha\alpha} - F'''_{\alpha\alpha}} = \frac{-(F_\alpha - F''_{\alpha\alpha})(\alpha_{13})_{\alpha_{12}} - H'_{r\alpha}(r_1)_{\alpha_{12}}}{1 + (\alpha_{14})_{\alpha_{12}}} \tag{VIII-7}$$

$$(F_\alpha - F''_{\alpha\alpha}) = \frac{-4(F_\alpha - F''''_{\alpha\alpha})(\alpha_{14})_{\alpha_{12}} - 4(\overline{F'_{\alpha\alpha} - F'''_{\alpha\alpha}})(\alpha_{13})_{\alpha_{12}} - 2H_{r\alpha}(r_1)_{\alpha_{12}}}{1 - (\alpha_{14})_{\alpha_{12}}} \tag{VIII-8}$$

In the above expressions, the terms such as $(r_1)_{\alpha_{12}}$ mean the displacement of bond distance r_1 which will minimize the energy after small unit positive displacement of angle α_{12}.

There are six genuine vibrations for octahedral MX_6 ($A_{1g} + E_g + 2F_{1u} + F_{2g} + F_{2u}$) as well as three redundant symmetry coordinates [$A_{1g} + E_g$ (a and b)]. Following a procedure similar to that outlined for square planar $M(CN)_4$ in Appendix II, we arrive at the F and G matrices given below.

$$
\begin{array}{lll}
A_{1g} & F_{11} = F_r + 4F''_{rr} + F'_{rr} & G_{11} = \mu_x \\
E_g & F_{22} = F_r - 2F''_{rr} + F'_{rr} & G_{22} = \mu_x \\
F_{1u} & F_{33} = F_r - F'_{rr} & G_{33} = 2\mu_M + \mu_x \\
 & F_{44} = (F_\alpha - F''_{\alpha\alpha}) + 2(\bar{F}'_{\alpha\alpha} - F'''_{\alpha\alpha}) & G_{44} = (8\mu_M + 2\mu_x)\rho^2 \\
 & F_{34} = 2(F'_{r\alpha} - F''_{r\alpha}) = 2H'_{r\alpha} & G_{34} = 4\mu_M\rho \\
F_{2g} & F_{55} = 2(F_\alpha - F''''_{\alpha\alpha}) - (F_\alpha - F''_{\alpha\alpha}) & G_{55} = 4\mu_x\rho \\
F_{2u} & F_{66} = (F_\alpha - F''_{\alpha\alpha}) - 2(\bar{F}'_{\alpha\alpha} - F'''_{\alpha\alpha}) & G_{66} = 2\mu_x\rho
\end{array}
$$

The internal coordinate force constants are defined by the expression (VIII-1) and Fig. VIII-1. As usual, μ_i is the reciprocal of mass of atom i and ρ is the reciprocal of equilibrium MX bond length.

Owing to the α redundancies, the force constants involving α appear only as linear combinations which cannot be separated. In fact they have no meaning by themselves. As pointed out in Sect. 1.13, compliance constants do not suffer from this fate; they can be determined independently. The symmetry compliance constants can be written as functions of the internal compliance constants exactly analogous to the relations above for force constants. However, the expressions can be simplified, since the symmetry compliance constant for a redundant coordinate is zero, leading to the relations

$$C''''_{\alpha\alpha} = -\tfrac{1}{2}(C_\alpha + C''_{\alpha\alpha}),\ C'''_{\alpha\alpha} = -C'_{\alpha\alpha}, \text{ and } C'_{\alpha r} = -C''_{\alpha r}.$$

Appendix IX

Character Tables for Various Point Groups

A few remarks about the following character tables will aid in their use. The upper left-hand corner gives the point group. The column headings give the different classes of symmetry operations and the number in each class. Thus, for C_{3v} there are 3 different vertical planes of symmetry (in the same class). The row label designates the irreducible representation. The numbers in the tables are the characters (sums of the diagonals of the transformation matrices). The last column shows what representation the quantities x^2, xy, etc. fall under. They indicate the *Raman activity as the polarizability derivative components* α_{xx}, α_{xy}, *etc., transform as the quadratics* x^2, xy, *etc.* An entry such as $x^2 + y^2 - 2z^2$ indicates components for α_{xx}, α_{yy}, and α_{zz}. The column second from right tells us under what representation the Cartesian coordinates transform. This indicates the *infrared activity as the changing dipole moment transforms as one or more of the Cartesian coordinates.* For those point groups which have a unique axis, it is designated as the Cartesian z axis.

It will be noted that for many of the point groups, such as C_3, C_5, C_6, etc., in the degenerate representations sometimes the symbol ϵ^k or ϵ^{k*} is listed. The symbol ϵ stands for $\exp(2\pi i/n)$ for an n-fold unique axis. The exponent k is an integer 1, 2, or 3. The asterisk * indicates the complex conjugate. It is to be noted that the symbol ϵ appears only in doubly degenerate representations for which two sets of characters are given. For applications to most physical problems, such as the symmetry classifications of molecular vibrations, it is appropriate to add the two sets of characters to obtain one set of characters to apply for a given problem. Thus for the characters of the E representation of C_3 we can use 2, -1, -1 instead of

$$\begin{Bmatrix} 1 & \epsilon & \epsilon^* \\ 1 & \epsilon^* & \epsilon \end{Bmatrix}$$

Some of the symmetry operations are ambiguous; the following comments are intended to clarify the ambiguity: C_n^m represents a clockwise

TABLE IX-1

C_s, C_i, and C_n

C_s	E	σ_h		
A'	1	1	x, y, R_z	x^2, y^2, z^2, xy
A''	1	-1	z, R_x, R_y	yz, xz

C_i	E	i		
A_g	1	1	R_x, R_y, R_z	$x^2, y^2, z^2, xy, xz, yz$
A_u	1	-1	x, y, z	

C_2	E	C_2		
A	1	1	z, R_z	x^2, y^2, z^2, xy
B	1	-1	x, y, R_x, R_y	yz, xz

C_3	E	C_3	C_3^2	$\epsilon = \exp(2\pi i/3)$; $\epsilon + \epsilon^* = -1$	
A	1	1	1	z, R_z	$x^2 + y^2, z^2,$ $x^2 - y^2, xy$
E	$\begin{Bmatrix} 1 \\ 1 \end{Bmatrix}$	$\begin{matrix} \epsilon \\ \epsilon^* \end{matrix}$	$\begin{matrix} \epsilon^* \\ \epsilon \end{matrix}$	x, y, R_x, R_y	xz, yz

C_4	E	C_4	C_2	C_4^3		
A	1	1	1	1	$z;\ R_z$	$x^2 + y^2, z^2$
B	1	-1	1	-1		$x^2 - y^2, xy$
E	$\begin{Bmatrix} 1 \\ 1 \end{Bmatrix}$	$\begin{matrix} i \\ -i \end{matrix}$	$\begin{matrix} -1 \\ -1 \end{matrix}$	$\begin{matrix} -i \\ i \end{matrix}$	$x, y;\ R_x, R_y$	xz, yz

C_5	E	C_5	C_5^2	C_5^3	C_5^4	$\epsilon = \exp(2\pi i/5)$; $\epsilon^n + \epsilon^{n*} = 2\cos(2\pi n/5)$	
A	1	1	1	1	1	$z;\ R_z$	$x^2 + y^2, z^2$
E_1	$\begin{Bmatrix} 1 \\ 1 \end{Bmatrix}$	$\begin{matrix} \epsilon \\ \epsilon^* \end{matrix}$	$\begin{matrix} \epsilon^2 \\ \epsilon^{2*} \end{matrix}$	$\begin{matrix} \epsilon^{2*} \\ \epsilon^2 \end{matrix}$	$\begin{matrix} \epsilon^* \\ \epsilon \end{matrix}$	$x, y;\ R_xR_y$	xz, yz
E_2	$\begin{Bmatrix} 1 \\ 1 \end{Bmatrix}$	$\begin{matrix} \epsilon^2 \\ \epsilon^{2*} \end{matrix}$	$\begin{matrix} \epsilon^* \\ \epsilon \end{matrix}$	$\begin{matrix} \epsilon \\ \epsilon^* \end{matrix}$	$\begin{matrix} \epsilon^{2*} \\ \epsilon^2 \end{matrix}$		$x^2 - y^2, xy$

TABLE IX-1 (contd.)

C_6	E	C_6	C_3	C_2	C_3^2	C_6^5	$\epsilon = \exp(2\pi i/6);\ \epsilon + \epsilon^* = 1$	
A	1	1	1	1	1	1	$z;\ R_z$	$x^2 + y^2, z^2$
B	1	−1	1	−1	1	−1		
E_1	1	ϵ	$-\epsilon^*$	−1	$-\epsilon$	ϵ^*	$x, y;\ R_x, R_y$	xz, yz
	1	ϵ^*	$-\epsilon$	−1	$-\epsilon^*$	ϵ		
E_2	1	$-\epsilon^*$	$-\epsilon$	1	$-\epsilon^*$	$-\epsilon$		$x^2 - y^2, xy$
	1	$-\epsilon$	$-\epsilon^*$	1	$-\epsilon$	$-\epsilon^*$		

C_7	E	C_7	C_7^2	C_7^3	C_7^4	C_7^5	C_7^6	$\epsilon = \exp(2\pi i/7);\ \epsilon^n + \epsilon^{n*} = 2\cos(2\pi n/7)$	
A	1	1	1	1	1	1	1	$z;\ R_z$	$x^2 + y^2, z^2$
E_1	1	ϵ	ϵ^2	ϵ^3	ϵ^{3*}	ϵ^{2*}	ϵ^*	$x, y:\ R_x, R_y$	xz, yz
	1	ϵ^*	ϵ^{2*}	ϵ^{3*}	ϵ^3	ϵ^2	ϵ		
E_2	1	ϵ^2	ϵ^{3*}	ϵ^*	ϵ	ϵ^3	ϵ^{2*}		$x^2 - y^2, xy$
	1	ϵ^{2*}	ϵ^3	ϵ	ϵ^*	ϵ^{3*}	ϵ^2		
E_3	1	ϵ^3	ϵ^*	ϵ^2	ϵ^{2*}	ϵ	ϵ^{3*}		
	1	ϵ^{3*}	ϵ	ϵ^{2*}	ϵ^2	ϵ^*	ϵ^3		

C_8	E	C_8	C_4	C_2	C_4^3	C_8^3	C_8^5	C_8^7	$\epsilon = \exp(2\pi i/8);\ \epsilon + \epsilon^* = \sqrt{2}$	
A	1	1	1	1	1	1	1	1	z, R_z	$x^2 + y^2, z^2$
B	1	−1	1	1	1	−1	−1	−1		
E_1	1	ϵ	i	−1	$-i$	$-\epsilon^*$	$-\epsilon$	ϵ^*	$x, y;\ R_x, R_y$	xz, yz
	1	ϵ^*	$-i$	−1	i	$-\epsilon$	$-\epsilon^*$	ϵ		
E_2	1	i	−1	1	−1	$-i$	i	$-i$		$x^2 - y^2, xy$
	1	$-i$	−1	1	−1	i	$-i$	i		
E_3	1	$-\epsilon$	i	−1	$-i$	ϵ^*	ϵ	$-\epsilon^*$		
	1	$-\epsilon^*$	$-i$	−1	i	ϵ	ϵ^*	$-\epsilon$		

TABLE IX-2

THE DIHEDRAL GROUP, D_n

D_2	E	$C_2(x)$	$C_2(y)$	$C_2(z)$		
A	1	1	1	1		x^2, y^2, z^2
B_1	1	1	−1	−1	$z;\ R_z$	xy
B_2	1	−1	1	−1	$y;\ R_y$	xz
B_3	1	−1	−1	1	$x;\ R_x$	yz

D_3	E	$2C_3$	$3C_2$		
A_1	1	1	1		$x^2 + y^2, z^2$
A_2	1	1	−1	$z;\ R_z$	
E	2	−1	0	$x, y;\ R_x, R_y$	$x^2 - y^2, xy, xz, yz$

D_4	E	$2C_4$	$C_4^2 = C_2$	$2C_2'$	$2C_2''$		
A_1	1	1	1	1	1		$x^2 + y^2, z^2$
A_2	1	1	1	−1	−1	$z; R_z$	
B_1	1	−1	1	1	−1		$x^2 - y^2$
B_2	1	−1	1	−1	1		xy
E	2	0	−2	0	0	$x, y;\ R_xR_y$	xz, yz

D_5	E	$2C_5$	$2C_5^2$	$5C_2$		
A_1	1	1	1	1		$x^2 + y^2, z^2$
A_2	1	1	1	−1	$z;\ R_z$	
E_1	2	2 cos 72°	2 cos 144°	0	$x, y;\ R_x, R_y$	xz, yz
E_2	2	2 cos 144°	2 cos 72°	0		$x^2 - y^2, xy$

D_6	E	$2C_6$	$2C_3$	C_2	$3C_2'$	$3C_2''$		
A_1	1	1	1	1	1	1		$x^2 + y^2, z^2$
A_2	1	1	1	1	−1	−1	$z;\ R_z$	
B_1	1	−1	1	−1	1	−1		
B_2	1	−1	1	−1	−1	1		
E_1	2	1	−1	−2	0	0	$x, y;\ R_x, R_y$	xz, yz
E_2	2	−1	−1	2	0	0		$x^2 - y^2, xy$

TABLE IX-3

C_{nv}

C_{2v}	E	C_2	$\sigma_v(xz)$	$\sigma_v(yz)$		
A_1	1	1	1	1	z	x^2, y^2, z^2
A_2	1	1	−1	−1	R_z	xy
B_1	1	−1	1	−1	$x;\ R_y$	xz
B_2	1	−1	−1	1	$y;\ R_x$	yz

C_{3v}	E	$2C_3$	$3\sigma_v$		
A_1	1	1	1	z	$x^2 + y^2, z^2$
A_2	1	1	−1	R_z	
E	2	−1	0	$x, y;\ R_xR_y$	$x^2 - y^2, xy, xz, yz$

C_{4v}	E	$2C_4$	C_2	$2\sigma_v$	$2\sigma_d$		
A_1	1	1	1	1	1	z	$x^2 + y^2, z^2$
A_2	1	1	1	−1	−1	R_z	
B_1	1	−1	1	1	−1		$x^2 - y^2$
B_2	1	−1	1	−1	1		xy
E	2	0	−2	0	0	$x, y;\ R_xR_y$	xz, yz

C_{5v}	E	$2C_5$	$2C_5^2$	$5\sigma_v$		
A_1	1	1	1	1	z	$x^2 + y^2, z^2$
A_2	1	1	1	−1	R_z	
E_1	2	2 cos 72°	2 cos 144°	0	$x, y;\ R_x, R_y$	xz, yz
E_2	2	2 cos 144°	2 cos 72°	0		$x^2 - y^2, xy$

C_{6v}	E	$2C_6$	$2C_3$	C_2	$3\sigma_v$	$3\sigma_d$		
A_1	1	1	1	1	1	1	z	$x^2 + y^2, z^2$
A_2	1	1	1	1	−1	−1	R_z	
B_1	1	−1	1	−1	1	−1		
B_2	1	−1	1	−1	−1	1		
E_1	2	1	−1	−2	0	0	$x, y;\ R_xR_y$	xz, yz
E_2	2	−1	−1	2	0	0		$x^2 - y^2, xy$

TABLE IX-4

C_{nh}

C_{2h}	E	C_2	i	σ_h		
A_g	1	1	1	1	R_z	x^2, y^2, z^2, xy
B_g	1	-1	1	-1	R_x, R_y	xz, yz
A_u	1	1	-1	-1	z	
B_u	1	-1	-1	1	x, y	

C_{3h}	E	C_3	C_3^2	σ_h	S_3	S_3^5	$\epsilon = \exp(2\pi i/3)$;	$\epsilon + \epsilon^* = -1$
A'	1	1	1	1	1	1	R_z	$x^2 + y^2, z^2$
E'	1	ϵ	ϵ^*	1	ϵ	ϵ^*	x, y	$x^2 - y^2, xy$
	1	ϵ^*	ϵ	1	ϵ^*	ϵ		
A''	1	1	1	-1	-1	-1	z	
E''	1	ϵ	ϵ^*	-1	$-\epsilon$	$-\epsilon^*$	R_x, R_y	xz, yz
	1	ϵ^*	ϵ	-1	$-\epsilon^*$	$-\epsilon$		

C_{4h}	E	C_4	C_2	C_4^3	i	S_4^3	σ_h	S_4		
A_g	1	1	1	1	1	1	1	1	R_z	$x^2 + y^2, z^2$
B_g	1	-1	1	-1	1	-1	1	-1		$x^2 - y^2, xy$
E_g	1	i	-1	$-i$	1	i	-1	$-i$	R_x, R_y	xz, yz
	1	$-i$	-1	i	1	$-i$	-1	i		
A_u	1	1	1	1	-1	-1	-1	-1	z	
B_u	1	-1	1	-1	-1	1	-1	1		
E_u	1	i	-1	$-i$	-1	$-i$	1	i	x, y	
	1	$-i$	-1	i	-1	i	1	$-i$		

C_{5h}	E	C_5	C_5^2	C_5^3	C_5^4	σ_h	S_5	S_5^7	S_5^3	S_5^9	$\varepsilon = \exp(2\pi i/5)$	
A'	1	1	1	1	1	1	1	1	1	1	R_z	$x^2 + y^2, z^2$
E_1'	1	ϵ	ϵ^2	ϵ^{2*}	ϵ^*	1	ϵ	ϵ^2	ϵ^{2*}	ϵ^*	x, y	
	1	ϵ^*	ϵ^{2*}	ϵ^2	ϵ	1	ϵ^*	ϵ^{2*}	ϵ^2	ϵ		
E_2'	1	ϵ^2	ϵ^*	ϵ	ϵ^{2*}	1	ϵ^2	ϵ^*	ϵ	ϵ^{2*}		$x^2 - y^2, xy$
	1	ϵ^{2*}	ϵ	ϵ^*	ϵ^2	1	ϵ^{2*}	ϵ	ϵ^*	ϵ^2		
A''	1	1	1	1	1	-1	-1	-1	-1	-1	z	
E_1''	1	ϵ	ϵ^2	ϵ^{2*}	ϵ^*	-1	$-\epsilon$	$-\epsilon^2$	$-\epsilon^{2*}$	$-\epsilon^*$	R_x, R_y	xz, yz
	1	ϵ^*	ϵ^{2*}	ϵ^2	ϵ	-1	$-\epsilon^*$	$-\epsilon^{2*}$	$-\epsilon^2$	$-\epsilon$		
E_2''	1	ϵ^2	ϵ^*	ϵ	ϵ^{2*}	-1	$-\epsilon^2$	$-\epsilon^*$	$-\epsilon$	$-\epsilon^{2*}$		
	1	ϵ^{2*}	ϵ	ϵ^*	ϵ^2	-1	$-\epsilon^{2*}$	$-\epsilon$	$-\epsilon^*$	$-\epsilon^2$		

C_{6h}	E	C_6	C_3	C_2	C_3^2	C_6^5	i	S_3^5	S_6^5	σ_h	S_6	S_3	$\epsilon = \exp(2\pi i/6)$; $\epsilon + \epsilon^* = 1$	
A_g	1	1	1	1	1	1	1	1	1	1	1	1	R_z	$x^2 + y^2, z^2$
B_g	1	-1	1	-1	1	-1	1	-1	1	-1	1	-1		
E_{1g}	1	ϵ	$-\epsilon^*$	-1	$-\epsilon$	ϵ^*	1	ϵ	$-\epsilon^*$	-1	$-\epsilon$	ϵ^*	(R_x, R_y)	xz, yz
	1	ϵ^*	$-\epsilon$	-1	$-\epsilon^*$	ϵ	1	ϵ^*	$-\epsilon$	-1	$-\epsilon^*$	ϵ		
E_{2g}	1	$-\epsilon^*$	$-\epsilon$	1	$-\epsilon^*$	$-\epsilon$	1	$-\epsilon^*$	$-\epsilon$	1	$-\epsilon^*$	$-\epsilon$		$x^2 - y^2, xy$
	1	$-\epsilon$	$-\epsilon^*$	1	$-\epsilon$	$-\epsilon^*$	1	$-\epsilon$	$-\epsilon^*$	1	$-\epsilon$	$-\epsilon^*$		
A_u	1	1	1	1	1	1	-1	-1	-1	-1	-1	-1	z	
B_u	1	-1	1	-1	1	-1	-1	1	-1	1	-1	1		
E_{1u}	1	ϵ	$-\epsilon^*$	-1	$-\epsilon$	ϵ^*	-1	$-\epsilon$	ϵ^*	1	ϵ	$-\epsilon^*$	x, y	
	1	ϵ^*	$-\epsilon$	-1	$-\epsilon^*$	ϵ	-1	$-\epsilon^*$	ϵ	1	ϵ^*	$-\epsilon$		
E_{2u}	1	$-\epsilon^*$	$-\epsilon$	1	$-\epsilon^*$	$-\epsilon$	-1	ϵ^*	ϵ	-1	ϵ^*	ϵ		
	1	$-\epsilon$	$-\epsilon^*$	1	$-\epsilon$	$-\epsilon^*$	-1	ϵ	ϵ^*	-1	ϵ	ϵ^*		

TABLE IX-5

D_{nh}

$D_{2h} = V_h$	E	$C_2(z)$	$C_2(y)$	$C_2(x)$	i	$\sigma(xy)$	$\sigma(zx)$	$\sigma(yz)$		
A_g	1	1	1	1	1	1	1	1		x^2, y^2, z^2
B_{1g}	1	1	−1	−1	1	1	−1	−1	R_z	xy
B_{2g}	1	−1	1	−1	1	−1	1	−1	R_y	xz
B_{3g}	1	−1	−1	1	1	−1	−1	1	R_x	yz
A_u	1	1	1	1	−1	−1	−1	−1		
B_{1u}	1	1	−1	−1	−1	−1	1	1	z	
B_{2u}	1	−1	1	−1	−1	1	−1	1	y	
B_{3u}	1	−1	−1	1	−1	1	1	−1	x	

D_{3h}	E	$2C_3$	$3C_2$	σ_h	$2S_3$	$3\sigma_v$		
A_1'	1	1	1	1	1	1		$x^2 + y^2, z^2$
A_2'	1	1	−1	1	1	−1	R_z	
E'	2	−1	0	2	−1	0	x, y	$x^2 - y^2, xy$
A_1''	1	1	1	−1	−1	−1		
A_2''	1	1	−1	−1	−1	1	z	
E''	2	−1	0	−2	1	0	R_x, R_y	xz, yz

D_{4h}	E	$2C_4$	C_2	$2C_2'$	$2C_2''$	i	$2S_4$	σ_h	$2\sigma_v$	$2\sigma_d$		
A_{1g}	1	1	1	1	1	1	1	1	1	1		$x^2 + y^2, z^2$
A_{2g}	1	1	1	−1	−1	1	1	1	−1	−1	R_z	
B_{1g}	1	−1	1	1	−1	1	−1	1	1	−1		$x^2 - y^2$
B_{2g}	1	−1	1	−1	1	1	−1	1	−1	1		xy
E_g	2	0	−2	0	0	2	0	−2	0	0	R_x, R_y	xz, yz
A_{1u}	1	1	1	1	1	−1	−1	−1	−1	−1		
A_{2u}	1	1	1	−1	−1	−1	−1	−1	1	1	z	
B_{1u}	1	−1	1	1	−1	−1	1	−1	−1	1		
B_{2u}	1	−1	1	−1	1	−1	1	−1	1	−1		
E_u	2	0	−2	0	0	−2	0	2	0	0	x, y	

D_{5h}	E	$2C_5$	$2C_5^2$	$5C_2$	σ_h	$2S_5$	$2S_5^3$	$5\sigma_v$		
A_1'	1	1	1	1	1	1	1	1		$x^2 + y^2, z^2$
A_2'	1	1	1	−1	1	1	1	−1	R_z	
E_1'	2	2 cos 72°	2 cos 144°	0	2	2 cos 72°	2 cos 144°	0	x, y	
E_2'	2	2 cos 144°	2 cos 72°	0	2	2 cos 144°	2 cos 72°	0		$x^2 - y^2, xy$
A_1''	1	1	1	1	−1	−1	−1	−1		
A_2''	1	1	1	−1	−1	−1	−1	1	z	
E_1''	2	2 cos 72°	2 cos 144°	0	−2	−2 cos 72°	−2 cos 144°	0	R_x, R_y	$xz \quad yz$
E_2''	2	2 cos 144°	2 cos 72°	0	−2	−2 cos 144°	−2 cos 72°	0		

TABLE IX-5 (contd.)

D_{6h}	E	$2C_6$	$2C_3$	C_2	$3C_2'$	$3C_2''$	i	$2S_3$	$2S_6$	σ_h	$3\sigma_d$	$3\sigma_v$		
A_{1g}	1	1	1	1	1	1	1	1	1	1	1	1		x^2+y^2, z^2
A_{2g}	1	1	1	1	−1	−1	1	1	1	1	−1	−1	R_z	
B_{1g}	1	−1	1	−1	1	−1	1	−1	1	−1	1	−1		
B_{2g}	1	−1	1	−1	−1	1	1	−1	1	−1	−1	1		
E_{1g}	2	1	−1	−2	0	0	2	1	−1	−2	0	0	R_x, R_y	xz, yz
E_{2g}	2	−1	−1	2	0	0	2	−1	−1	2	0	0		x^2-y^2, xy
A_{1u}	1	1	1	1	1	1	−1	−1	−1	−1	−1	−1		
A_{2u}	1	1	1	1	−1	−1	−1	−1	−1	−1	1	1	z	
B_{1u}	1	−1	1	−1	1	−1	−1	1	−1	1	−1	1		
B_{2u}	1	−1	1	−1	−1	1	−1	1	−1	1	1	−1		
E_{1u}	2	1	−1	−2	0	0	−2	−1	1	2	0	0	x, y	
E_{2u}	2	−1	−1	2	0	0	−2	1	1	−2	0	0		

TABLE IX-6

D_{nd}

$D_{2d}=V_d$	E	$2S_4$	C_2	$2C_2'$	$2\sigma_d$		
A_1	1	1	1	1	1		x^2+y^2, z^2
A_2	1	1	1	−1	−1	R_z	
B_1	1	−1	1	1	−1		x^2-y^2
B_2	1	−1	1	−1	1	z	xy
E	2	0	−2	0	0	$x, y;\ R_x, R_y$	xz, yz

D_{3d}	E	$2C_3$	$3C_2$	i	$2S_6$	$3\sigma_d$		
A_{1g}	1	1	1	1	1	1		x^2+y^2, z^2
A_{2g}	1	1	−1	1	1	−1	R_z	
E_g	2	−1	0	2	−1	0	R_x, R_y	x^2-y^2, xy, xz, yz
A_{1u}	1	1	1	−1	−1	−1		
A_{2u}	1	1	−1	−1	−1	1	z	
E_u	2	−1	0	−2	1	0	x, y	

D_{4d}	E	$2S_8$	$2C_4$	$2S_8^3$	C_2	$4C_2'$	$4\sigma_d$		
A_1	1	1	1	1	1	1	1		x^2+y^2, z^2
A_2	1	1	1	1	1	−1	−1	R_z	
B_1	1	−1	1	−1	1	1	−1		
B_2	1	−1	1	−1	1	−1	1	z	
E_1	2	$\sqrt{2}$	0	$-\sqrt{2}$	−2	0	0	x, y	
E_2	2	0	−2	0	2	0	0		x^2-y^2, xy
E_3	2	$-\sqrt{2}$	0	$\sqrt{2}$	−2	0	0	R_x, R_y	xz, yz

TABLE IX-6 (contd.)

D_{5d}	E	$2C_5$	$2C_5^2$	$5C_2$	i	$2S_{10}^3$	$2S_{10}$	$5\sigma_d$		
A_{1g}	1	1	1	1	1	1	1	1		x^2+y^2, z^2
A_{2g}	1	1	1	−1	1	1	1	−1	R_z	
E_{1g}	2	2 cos 72°	2 cos 144°	0	2	2 cos 72°	2 cos 144°	0	R_x, R_y	xz, yz
E_{2g}	2	2 cos 144°	2 cos 72°	0	2	2 cos 144°	2 cos 72°	0		x^2-y^2, xy
A_{1u}	1	1	1	1	−1	−1	−1	−1		
A_{2u}	1	1	1	−1	−1	−1	−1	1	z	
E_{1u}	2	2 cos 72°	2 cos 144°	0	−2	−2 cos 72°	−2 cos 144°	0	x, y	
E_{2u}	2	2 cos 144°	2 cos 72°	0	−2	−2 cos 144°	−2 cos 72°	0		

D_{6d}	E	$2S_{12}$	$2C_6$	$2S_4$	$2C_3$	$2S_{12}^5$	C_2	$6C_2'$	$6\sigma_d$		
A_1	1	1	1	1	1	1	1	1	1		x^2+y^2, z^2
A_2	1	1	1	1	1	1	1	−1	−1	R_z	
B_1	1	−1	1	−1	1	−1	1	1	−1		
B_2	1	−1	1	−1	1	−1	1	−1	1	z	
E_1	2	$\sqrt{3}$	1	0	−1	$-\sqrt{3}$	−2	0	0	x, y	
E_2	2	1	−1	−2	−1	1	2	0	0		x^2-y^2, xy
E_3	2	0	−2	0	2	0	−2	0	0		
E_4	2	−1	−1	2	−1	−1	2	0	0		
E_5	2	$-\sqrt{3}$	1	0	−1	$\sqrt{3}$	−2	0	0	R_x, R_y	xz, yz

TABLE IX-7

S_n

S_4	E	S_4	C_2	S_4^3		
A	1	1	1	1	R_z	x^2+y^2, z^2
B	1	−1	1	−1	z	x^2-y^2, xy
E	$\begin{Bmatrix}1 & i & -1 & -i \\ 1 & -i & -1 & i\end{Bmatrix}$				$x, y;\ R_x, R_y$	xz, yz

S_6	E	C_3	C_3^2	i	S_6^5	S_6	$\epsilon = \exp(2\pi i/3);\ \epsilon + \epsilon^* = -1$	
A_g	1	1	1	1	1	1	R_z	x^2+y^2, z^2
E_g	$\begin{Bmatrix}1 & \epsilon & \epsilon^* & 1 & \epsilon & \epsilon^* \\ 1 & \epsilon^* & \epsilon & 1 & \epsilon^* & \epsilon\end{Bmatrix}$						R_x, R_y	x^2-y^2, xy, xz, yz
A_u	1	1	1	−1	−1	−1	z	
E_u	$\begin{Bmatrix}1 & \epsilon & \epsilon^* & -1 & -\epsilon & -\epsilon^* \\ 1 & \epsilon^* & \epsilon & -1 & -\epsilon^* & -\epsilon\end{Bmatrix}$						x, y	

S_8	E	S_8	C_4	S_8^3	C_2	S_8^5	C_4^3	S_8^7	$\epsilon = \exp(2\pi i/8);\ \epsilon + \epsilon^* = \sqrt{2}$	
A	1	1	1	1	1	1	1	1	R_z	x^2+y^2, z^2
B	1	−1	1	−1	1	−1	1	−1	z	
E_1	$\begin{Bmatrix}1 & \epsilon & i & -\epsilon^* & -1 & -\epsilon & -i & \epsilon^* \\ 1 & \epsilon^* & -i & -\epsilon & -1 & -\epsilon^* & i & \epsilon\end{Bmatrix}$								$x, y;\ R_y, R_y$	
E_2	$\begin{Bmatrix}1 & i & -1 & -i & 1 & i & -1 & -i \\ 1 & -i & -1 & i & 1 & -i & -1 & i\end{Bmatrix}$									x^2-y^2, xy
E_3	$\begin{Bmatrix}1 & -\epsilon^* & -i & \epsilon & -1 & \epsilon^* & i & -\epsilon \\ 1 & -\epsilon & i & \epsilon^* & -1 & \epsilon & -i & -\epsilon^*\end{Bmatrix}$									xz, yz

TABLE IX-8

THE CUBIC GROUPS

T	E	$4C_3$	$4C_3^2$	$3C_2$		$\epsilon = \exp(2\pi i/3); \epsilon + \epsilon^* = -1$
A	1	1	1	1		$x^2 + y^2 + z^2$
E	1	ϵ	ϵ^*	1		$x^2 + y^2 - 2z^2;\ x^2 - y^2$
	1	ϵ^*	ϵ	1		
F	3	0	0	-1	$x, y, z,$ R_x, R_y, R_z	xy, xz, yz

T_h	E	$4C_3$	$4C_3^2$	$3C_2$	i	$4S_6$	$4S_6^5$	$3\sigma_h$		$\epsilon = \exp(2\pi i/3); \epsilon + \epsilon^* = -1$
A_g	1	1	1	1	1	1	1	1		$x^2 + y^2 + z^2$
E_g	1	ϵ	ϵ^*	1	1	ϵ	ϵ^*	1		$x^2 + y^2 - 2z^2, x^2 - y^2$
	1	ϵ^*	ϵ	1	1	ϵ^*	ϵ	1		
F_g	3	0	0	-1	3	0	0	-1	R_x, R_y, R_z	xy, xz, yz
A_u	1	1	1	1	-1	-1	-1	-1		
E_u	1	ϵ	ϵ^*	1	-1	$-\epsilon$	$-\epsilon^*$	-1		
	1	ϵ^*	ϵ	1	-1	$-\epsilon^*$	$-\epsilon$	-1		
F_u	3	0	0	-1	-3	0	0	1	x, y, z	

T_d	E	$8C_3$	$3C_2$	$6S_4$	$6\sigma_d$		
A_1	1	1	1	1	1		$x^2 + y^2 + z^2$
A_2	1	1	1	-1	-1		
E	2	-1	2	0	0		$x^2 + y^2 - 2z^2, x^2 - y^2$
F_1	3	0	-1	1	-1	R_x, R_y, R_z	
F_2	3	0	-1	-1	1	x, y, z	xy, xz, yz

O	E	$8C_3$	$3C_2$	$6C_4$	$6C_2'$		
A_1	1	1	1	1	1		$x^2 + y^2 + z^2$
A_2	1	1	1	-1	-1		
E	2	-1	2	0	0		$x^2 + y^2 - 2z^2, x^2 - y^2$
F_1	3	0	-1	1	-1	R_x, R_y, R_z, x, y, z	
F_2	3	0	-1	-1	1		xy, xz, yz

O_h	E	$8C_2$	$6C_2$	$6C_4$	$3C_2 (= C_4^2)$	i	$6S_4$	$8S_6$	$3\sigma_h$	$6\sigma_d$		
A_{1g}	1	1	1	1	1	1	1	1	1	1		$x^2 + y^2 + z^2$
A_{2g}	1	1	-1	-1	1	1	-1	1	1	-1		
E_g	2	-1	0	0	2	2	0	-1	2	0		$x^2 + y^2 - 2z^2, x^2 - y^2$
F_{1g}	3	0	-1	1	-1	3	1	0	-1	-1	R_x, R_y, R_z	
F_{2g}	3	0	1	-1	-1	3	-1	0	-1	1		xy, xz, yz
A_{1u}	1	1	1	1	1	-1	-1	-1	-1	-1		
A_{2u}	1	1	-1	-1	1	-1	1	-1	-1	1		
E_u	2	-1	0	0	2	-2	0	1	-2	0		
F_{1u}	3	0	-1	1	-1	-3	-1	0	1	1	x, y, z	
F_{2u}	3	0	1	-1	-1	-3	1	0	1	-1		

TABLE IX-8 (contd.)

I	E	$12C_5$	$12C_5^2$	$20C_3$	$15C_2$		
A	1	1	1	1	1		$x^2+y^2+z^2$
F_1	3	$\frac{1+\sqrt{5}}{2}$	$\frac{1-\sqrt{5}}{2}$	0	−1	$\begin{Bmatrix} x, y, z; \\ R_x, R_y, R_z \end{Bmatrix}$	
F_2	3	$\frac{1-\sqrt{5}}{2}$	$\frac{1+\sqrt{5}}{2}$	0	−1		
G	4	−1	−1	1	0		
H	5	0	0	−1	1		$\begin{Bmatrix} x^2+y^2-2z^2, \\ x^2-y^2, xy, xz, yz \end{Bmatrix}$

I_h	E	$12C_5$	$12C_5^2$	$20C_3$	$15C_2$	i	$12S_{10}$	$12S_{10}^3$	$20S_6$	15σ		
A_g	1	1	1	1	1	1	1	1	1	1		$x^2+y^2+z^2$
F_{1g}	3	$\frac{1+\sqrt{5}}{2}$	$\frac{1-\sqrt{5}}{2}$	0	−1	3	$\frac{1-\sqrt{5}}{2}$	$\frac{1+\sqrt{5}}{2}$	0	−1	$\begin{Bmatrix} R_x, R_y, \\ R_z \end{Bmatrix}$	
F_{2g}	3	$\frac{1-\sqrt{5}}{2}$	$\frac{1+\sqrt{5}}{2}$	0	−1	3	$\frac{1+\sqrt{5}}{2}$	$\frac{1-\sqrt{5}}{2}$	0	−1		
G_g	4	−1	−1	1	0	4	−1	−1	1	0		
H_g	5	0	0	−1	1	5	0	0	−1	1		$\begin{Bmatrix} x^2+y^2-2z^2, \\ x^2-y^2, \\ xy, xz, yz \end{Bmatrix}$
A_u	1	1	1	1	1	−1	−1	−1	−1	−1		
F_{1u}	3	$\frac{1+\sqrt{5}}{2}$	$\frac{1-\sqrt{5}}{2}$	0	−1	3	$\frac{-(1-\sqrt{5})}{2}$	$\frac{-(1+\sqrt{5})}{2}$	0	1	x, y, z	
F_{2u}	3	$\frac{1-\sqrt{5}}{2}$	$\frac{1+\sqrt{5}}{2}$	0	−1	−3	$\frac{-(1+\sqrt{5})}{2}$	$\frac{-(1-\sqrt{5})}{2}$	0	1		
G_u	4	−1	−1	1	0	−4	1	1	−1	0		
H_u	5	0	0	−1	1	−5	0	0	1	−1		

TABLE IX-9

LINEAR GROUPS

$C_{\infty v}$	E	$2C_\infty^\varphi$	$\cdots$	$\infty\sigma_v$		
$A_1 \equiv \Sigma^+$	1	1	$\cdots$	1	z	x^2+y^2, z^2
$A_2 \equiv \Sigma^-$	1	1	$\cdots$	−1	R_z	
$E_1 \equiv \Pi$	2	$2\cos\varphi$	$\cdots$	0	$x, y;\ R_x, R_y$	xz, yz
$E_2 \equiv \Delta$	2	$2\cos 2\varphi$	$\cdots$	0		x^2-y^2, xy
$E_3 \equiv \varphi$	2	$2\cos 3\varphi$	$\cdots$	0		
$\cdots$	$\cdots$	$\cdots$	$\cdots$	$\cdots$		

TABLE IX-9 (contd.)

$D_{\infty h}$	E	$2C_\infty^\varphi$	$\cdots$	$\infty\sigma_v$	i	$2S_\infty^\varphi$	$\cdots$	∞C_2	σ_h		
Σ_g^+	1	1	$\cdots$	1	1	1	$\cdots$	1	1		$x^2 + y^2, z^2$
Σ_g^-	1	1	$\cdots$	−1	1	1	$\cdots$	−1	1	R_z	
Π_g	2	$2 \cos \varphi$	$\cdots$	0	2	$-2 \cos \varphi$	$\cdots$	0	−2	R_x, R_y	xz, yz
Δ_g	2	$2 \cos 2\varphi$	$\cdots$	0	2	$2 \cos 2\varphi$	$\cdots$	0	2		$x^2 - y^2, xy$
$\cdots$	$\cdots$	$\cdots$	$\cdots$	$\cdots$	$\cdots$	$\cdots$	$\cdots$	$\cdots$	$\cdots$		
Σ_u^+	1	1	$\cdots$	1	−1	−1	$\cdots$	−1	−1	z	
Σ_u^-	1	1	$\cdots$	−1	−1	−1	$\cdots$	1	−1		
Π_u	2	$2 \cos \varphi$	$\cdots$	0	−2	$2 \cos \varphi$	$\cdots$	0	2	x, y	
Δ_u	2	$2 \cos 2\varphi$	$\cdots$	0	−2	$-2 \cos 2\varphi$	$\cdots$	0	−2		
$\cdots$	$\cdots$	$\cdots$	$\cdots$	$\cdots$	$\cdots$	$\cdots$	$\cdots$	$\cdots$	$\cdots$		

rotation of $(2\pi m/n)$ about an n-fold axis. For D_2 and D_{2h} point groups, there is no unique axis, so the Cartesian labeling is arbitrary. There may be rotation axes of the same degree but different classes in a given point group. For example, for D_{4h}, such as planar XY_4, there are three different classes of two-fold axis. One is coincident with the C_4 axis ($C_2 = C_4^2$). Another is along a YXY axis (C_2'), while the third bisects the 90° YXY angles (C_2''). The choice of C_2' and C_2'' (and the accompanying representation nomenclature) is somewhat arbitrary; no trouble arises if one is consistent. For tables correlating point group symmetry elements with some of their subgroup symmetry elements, the reader is referred to Appendix X of *Molecular Vibrations* by E. B. Wilson, J. C. Decius, and P. C. Cross (*1*). This same reference is useful for determining representations of overtones and combination bands and for further details of character tables and selection rules.

REFERENCE

1. E. B. Wilson, J. C. Decius, and P. C. Cross, *Molecular Vibrations*, McGraw-Hill, New York, 1955.

Author Index

Numbers in parentheses are reference numbers and indicate that an author's work is referred to although his name is not cited in the text. Numbers in italics give the page on which the complete reference is listed.

Subject Index